ICT FUTURES

ICT FUTURES:
DELIVERING PERVASIVE, REAL-TIME AND SECURE SERVICES

Edited by

Paul Warren, John Davies and David Brown
All of BT, UK

John Wiley & Sons, Ltd

Other Wiley Editorial Offices

John Wiley & Sons Inc., 111 River Street, Hoboken, NJ 07030, USA

Jossey-Bass, 989 Market Street, San Francisco, CA 94103-1741, USA

Wiley-VCH Verlag GmbH, Boschstr. 12, D-69469 Weinheim, Germany

John Wiley & Sons Australia Ltd, 42 McDougall Street, Milton, Queensland 4064, Australia

John Wiley & Sons (Asia) Pte Ltd, 2 Clementi Loop #02-01, Jin Xing Distripark, Singapore 129809

John Wilcy & Sons Canada Ltd, 6045 Freemont Blvd, Mississauga, ONT, L5R 4J3, Canada

Wiley also publishes its books in a variety of electronic formats. Some content that appears in print may
not be available in electronic books.

Library of Congress Cataloging-in-Publication Data

ICT futures: delivering pervasive, real-time, and secure services / edited by Paul Warren, John Davies,
and David Brown.
        p.   cm.
    Includes bibliographical references and index.
    ISBN 978-0-470-99770-3 (cloth)
    1. Telecommunication–Technological innovations–Forecasting.   2. Computer networks–Security
measures–Forecasting.   3. Internet commerce–Forecasting.   4. Internet–Social aspects–
Forecasting.   5. Web sites–Forecasting.   I. Warren, Paul (Paul W.)   II. Davies, J. (N. John)
III. Brown, David, 1954–
    TK5102.5.I295   2008
    384–dc22

                                                                                    2008006097

British Library Cataloguing in Publication Data

A catalogue record for this book is available from the British Library

ISBN 978-0-470-99770-3 (H/B)

Typeset in 10/12pt Times by SNP Best-set Typesetter Ltd., Hong Kong
Printed and bound in Great Britain by Antony Rowe Ltd, Chippenham, England.

Contents

Part Four   Final Words

Editor Biographies

Paul Warren works within the IT Futures Centre at BT. He is currently Project Director of ACTIVE (http://www.active-project.eu), a European project in the area of collaborative knowledge management. Recent previous roles have been in the areas of service-oriented infrastructure and semantic technologies – both at the forefront of developments in ICT. At an earlier stage in his career, Paul worked on technology strategy and technology foresight, investigating areas as diverse as eBusiness and novel forms of computing. Paul has published widely on knowledge management, semantic technologies and technology foresight. Paul holds a degree in Theoretical Physics from Cambridge University and a MSc in Electronics from Southampton University. Paul can be contacted at paul.w.warren@bt.com

Dr John Davies leads the Next Generation Web research group within the IT Futures Centre at BT. Current interests centre around the application of semantic web and Web 2.0 technology to knowledge management, business intelligence, information integration and service-oriented environments. John is chairman of the European Semantic Technology Conference (www.estc2007.org). He has written and edited many papers and books in the area of semantic technology and its business applications, web-based information management and knowledge management; and has served on the program committee of numerous conferences in these areas. He is a Fellow of the British Computer Society and a Chartered Engineer. Earlier research at BT led to the development of a set of knowledge management tools which are the subject of a number of patents. These tools were spun out of BT and are now marketed by Infonic Ltd, of which John is Group Technical Advisor. John received the BT Technology Award for Technology Entrepreneurship for his contribution to the creation of Infonic. John can be contacted at john.nj.davies@bt.com

David Brown is Head of the Foresight team in BT Research and Venturing where his key role is to help BT and its customers manage the future. His role is particularly important to the development of BT, and along with the contribution of his other team members, he is successfully predicting fast-paced changes in technology. This ensures that BT invests wisely and continues to offer the right services needed by customers, both for today and tomorrow. David has written many articles on technology strategy and economics. Previous responsibility within his time at BT includes leading teams in the network strategy, product management and applications development fields. Before joining BT, David worked for ICL as a systems engineer. David gained a BSc in Mathematics from Durham University and holds a Sloan Fellowship MSc in Economics from the London Business School. David can be contacted at dave.g.brown@bt.com

List of Contributors

Ben Anderson
Technology and Social Change Research
  Centre
University of Essex,
Wivenhoe Park,
Colchester CO4 3SQ, UK
benander@essex.ac.uk

Sinan Aral
Department of Information, Operations
  & Management Sciences
NYU Stern School of Business,
Kaufman Management Center,
14 West 4th Street,
New York, NY, 10012, USA
sinan@stern.nyu.edu

Richard Benjamins
Telefonica R&D
Calle de Emilio Vargas,
28043, Madrid, Spain
rbenjamins@tid.es

Janette Bennett
BT
Fleet Place House, 2 Fleet Place,
London EC4M 7ET, UK
janette.bennett@bt.com

George Bilchev
BT
Adastral Park,
Martlesham Health,
Ipswich, Suffolk IP5 3RE, UK
george.bilchev@bt.com

Edgar E. Blanco
MIT Center for Transportation &
  Logistics
Massachusetts Institute of Technology,
77 Massachusetts Avenue,
Cambridge, MA 02139, USA
eblanco@mit.edu

Erik Brynjolfsson
Center for Digital Business
MIT Sloan School,
50 Memorial Drive,
Cambridge, MA 02142, USA
erikb@mit.edu

Chris Caplice
MIT Center for Transportation &
  Logistics
Massachusetts Institute of Technology,
1 Amherst Street,
Cambridge, MA 02139, USA
caplice@mit.edu

Pompeu Casanovas
Institut de Dret i Tecnologia UAB
UAB Institute of Law and
  Technology,
Political Science and Public Law
  Department,
Campus UAB,
Facultat de Dret,
08193 Bellaterra, Barcelona, Spain
pompeu.casanovas@uab.cat

Jon Clark
BT,
Adastral Park,
Martlesham Health,
Ipswich, Suffolk IP5 3RE, UK
jonathan.a.clark@bt.com

Mark Dames
BT
Alexander Graham Bell House,
1 Lochside View,
Edinburgh Park,
Edinburgh, Midlothian EH12 9DH, UK
mark.dames@bt.com

Richard Dennis
BT
Adastral Park,
Martlesham Health,
Ipswich, Suffolk IP5 3RE, UK
richard.dennis@bt.com

Theo Dimitrakos
BT
Adastral Park,
Martlesham Health,
Ipswich, Suffolk IP5 3RE, UK
theo.dimitrakos@bt.com

Ivan Djordjevic
CA
Ditton Park, Riding Court Road,
Datchet,
Slough SL3 9LL, UK
ivan.djordjevic@ca.com

Mike Fisher
BT
Adastral Park,
Martlesham Health,
Ipswich, Suffolk IP5 3RE, UK
mike.fisher@bt.com

Paul Garner
BT
Adastral Park,
Martlesham Health,
Ipswich, Suffolk IP5 3RE, UK
paul.2.garner@bt.com

Lesley Gavin
BT
81 Newgate Street,
London EC1A 7AJ, UK
lesley.gavin@bt.com

David Heatley
BT
Adastral Park,
Martlesham Health,
Ipswich, Suffolk IP5 3RE, UK
dave.heatley@bt.com

Nicholas J. Kings
BT
Adastral Park,
Martlesham Health,
Ipswich, Suffolk IP5 3RE, UK
nick.kings@bt.com

James McDonald
BT
Adastral Park,
Martlesham Health,
Ipswich, Suffolk IP5 3RE, UK
james.mcdonald@bt.com

Paul Mckee
BT
Adastral Park,
Martlesham Health,
Ipswich, Suffolk IP5 3RE, UK
paul.mckee@bt.com

Jonathan Mitchener
BT
Adastral Park,
Martlesham Health,
Ipswich, Suffolk IP5 3RE, UK
jonathan.mitchener@bt.com

David Payne
Institute of Advanced
    Telecommunications
Swansea University,
Singleton Park
Swansea SA2 8PP, UK
david.b.payne@btinternet.com

Ian Pearson
Futurizon GmbH
Buttikon, Switzerland
idpearson@gmail.com

Marta Poblet
Institute de Dret i Technologia
    UAB
UAB Institute of Law and
    Technology,
Political Science and Public Law
    Department,
Campus UAB, Facultat de Dret,
08193 Bellaterra. Barcelona. Spain.
marta.poblet@uab.es

Paul Stoneman
Chimera Institute
University of Essex,
Wivenhoe Park,
Colchester, Essex CO4 3SQ, UK
pstone@essex.ac.uk

Richard Tateson
BT
Adastral Park,
Martlesham Health,
Ipswich, Suffolk IP5 3RE, UK
richard.tateson@bt.com

Simon Thompson
BT
Adastral Park,
Martlesham Health,
Ipswich, Suffolk IP5 3RE, UK
simon.2.thompson@bt.com

Glen L. Urban
MIT Center for Digital Business
Cambridge Center,
NE20-336,
Cambridge, MA 02142, USA
glurban@mit.edu

Marshall van Alstyne
MIT – Center for E-Business
50 Memorial Drive,
Cambridge, MA 02139-4307, USA
marshall@mit.edu
    and
Boston University
595 Commonwealth Ave,
Boston, MA 02215, USA
mva@bu.edu

David Verrill
The MIT Center for Digital Business
3 Cambridge Center,
Cambridge, MA 02142, USA
dverrill@mit.edu

Dave Wisely
BT
Adastral Park,
Martlesham Health,
Ipswich, Suffolk IP5 3RE, UK
dave.wisely@bt.com

John Wittgreffe
BT
Adastral Park,
Martlesham Health,
Ipswich, Suffolk IP5 3RE, UK
john.wittgreffe@bt.com

Chris Wroe
BT
St Giles House, 1 Drury Lane,
Drury Lane,
London WC2B 5RS, UK
chris.wroe@bt.com

Foreword

The pace of technological change is relentless. It took the automobile industry a hundred years to penetrate 50 percent of the global market and 75 years for fixed line phones to reach 50 percent penetration. Look at the explosive growth in take-up of broadband and PC ownership. Even more astonishing is the adoption of cellular phones that has passed 2 billion in less than two decades – that's a mobile for every third person on this planet. Voice over the Internet and Web 2.0 applications, meanwhile, are already demonstrating even quicker adoption curves.

The past decade has seen tremendous progress in many different areas of information and communication technology (ICT). The increases in the power of microprocessors, the capacities of disc drives and storage arrays and the bandwidths of data networks are examples, but equally impressive is the progress in applying ICT across industries. In each case, performance has advanced at an exponential rate, making a whole lot more possible.

Now imagine each direction of progress of each technology as a vector whose length defines what is possible. Imagine further that the set of vectors is arranged as the radii of a sphere of possibilities. You can add to this new radii representing breakthrough technologies like broadband mobile, quantum computing, nano-technology and semantic web. This is where the combined impact of all the separate advances becomes obvious. If each single radius were to grow by a factor of 10, the sphere that defines what is possible would grow by 100000 percent!

The result is an explosion of creativity that shows no sign of coming to an end. I call this phenomenon the Innovation Big Bang. The rate of technology change is at a pace that has never been seen before. There was a lot of apparent innovation in the dot-com bubble but in hindsight much of it was bad cloning. Today is different because there is more fundamental and meaningful innovation happening in technology than at any time in history.

The ICT industry is undergoing profound change, whilst profoundly changing the world economic landscape. The most fundamental change however is the pervasiveness of ICT throughout all our business and leisure activities. For example, instead of having to chase around for information, information will reach you wherever you are, whenever you want it.

ICT driving innovation not at the speed of technology, but at the speed of our customers' lives – whether it's in their personal lives, whether it's in their professional lives or whether it's in the future state of their businesses. Nobody has a crystal ball on what customers want, and most of the time customers won't know exactly what they want – but they will always know good when they see it.

However we are heading for a world where we know a lot about the ingredients that will make up customer solutions – presence and messaging and authentication and content repurposing and digital rights management, storage and security – building software-based capabilities that allow us to separate the higher order brain functions from the infrastructure itself. We will be able to take these emerging capabilities and put them together to meet needs that we don't even know exist yet.

If you want to gain insights into the transformational impact ICT is having on the way we live and work, you will find this book engaging. It tackles the relationship between technology and people, and gives thoughtful examples of how technology is impacting innovation in areas as diverse as the retail supply chain and law.

Most importantly, it has been written by the people who are creating the next generation of ICT. Some are from my team in the Chief Technology Office in BT, some are our colleagues in other parts of BT, while others are collaboration partners from research organizations around the world. For BT, the amount of innovation we're capable of delivering is no longer defined only by the size of our R&D budget – it's as big as our global innovation network.

I'm convinced that this book will inspire people and enable them to use the insight it provides to build on and exploit the changing ICT landscape, in the spirit of open innovation.

ICT is already transforming businesses and everyday lives. It's the investment in an ICT future that will enable us to deliver new innovations throughout the globe faster. It will help our customers to respond to change as never before and enable deep collaboration across and beyond the enterprise. It will provide a consistently great customer experience with minimized environmental impact and the ability to operate securely and adapt quickly in an era where the intensity of competition is ever greater.

Matt Bross
Group Chief Technology Officer
BT

Introduction

Paul Warren, John Davies and David Brown

The Goal of Our Book

We have created this book precisely because it is a book we want to read. Finding no such book existed, our only recourse was to collaborate with colleagues to create one. We wanted to understand what we could expect of Information and Communication Technology (ICT) in the coming decade and how ICT would alter our private and professional lives. Specifically, we wanted that understanding encompassed within one single book. The intended audience for such a book would comprise, in part, people like ourselves: people working in the field and perhaps expert in some aspects of ICT, but of course not expert in all aspects – because no such person could exist. Equally importantly, our intended audience comprises those who do not regard themselves as primarily ICT professionals, but who need to understand how the technology is developing and the direction in which it is taking us all. Amongst such readers will, we hope, be managers responsible for decisions about the use of ICT in their organisations, and also innovative users who are seeking to get the most out of ICT in their professional and private lives.

We stress the timescale of approximately the next decade. With the exception of our penultimate chapter, this is not a book of futurology. Futurology is notoriously difficult. Our goals are less ambitious, but still challenging. We want to give a picture of what ICT will be like over a timescale which practically minded people need to begin thinking about now. We leave the reader to judge how well we have achieved this.

ICT and Society

Wikipedia defines ICT as 'Information Communication Technology, a broad subject concerned with technology and other aspects of managing and processing information'. This is good as a starting point. Certainly, the adjective 'broad' is well chosen. The 'T' might better stand for technologies, since we are clearly talking here about bringing together a range of quite different technologies. To 'managing and processing', we would want to add 'transmitting'. We would also want to say something – a very important something – about the interaction between people and information.

In essence, we are talking about the technologies which enable us to use information, for our individual and corporate gain. These are the technologies which have transformed our society in the last half decade. The precise quantitative economic effect of ICT is a matter for economic research. Certainly, a remark two decades ago by the economist and Nobel prize winner Robert Solow (1987), 'You can see the computer age everywhere but in the productivity statistics', is no longer credible – if it ever was. For example, a study of productivity growth and ICT in the UK for the period 1970–2000 (Oulton and Srinivasan, 2005) concluded that 'The accumulation of ICT capital has played an increasingly important, and in the second half of the 1990s the dominant, role in accounting for labour productivity growth in the market sector. ICT capital deepening accounted for 13% of the growth of output per hour in the market sector in 1970–79, 26% in 1979–90, and 28% in 1990–2000. In 1995–2000 the proportion rises to 47%'.

In any case, whatever we may think about the quantitative effect of ICT, the qualitative effect is beyond dispute. Our working life has changed: the way we access and use information, with instantaneous access to what used to take days or weeks to obtain; the kind of work we do, with routine clerical tasks automated; and where work is done, with the outsourcing of call centres and back-office functions to other continents. Our personal lives, too, have changed: with the internet used to buy and sell and also to interact with our friends, and on a darker note, with concerns about identity fraud and the misuse of confidential data.

The case for the impact of ICT on society is unanswerable. In all probability, future impact will be at least as great as we have seen so far. We attempt here to say something about that future impact.

A Preview

After this introduction, our book is divided into four parts. We begin with *People and Technology*, by looking at the relationship between technology and its users generally, and specifically at the points of interaction. We maintain that any consideration of technology which is not grounded in an understanding of how people interact with that technology is profoundly missing the point; any forecasts which do not take into account that interaction will inevitably be wrong.

In *Building the Infrastructure*, we adopt a more avowedly technological focus. The aim here is to describe the key components of ICT, how they are developing, and how they contribute to the whole.

In *Applying Technology*, we look at how technology is being used to create applications that benefit people. We take a sector-based approach. However, much of what is described is relevant outside the particular sectors described.

In our last part, *Final Words*, we first take a rather longer-term view and then in the concluding chapter we emphasise the most prominent trends which have run through our book.

People and Technology

Chapter 1 starts this part by talking about the nature of socio-technical prediction itself. It suggests that a simple model, in which predictable social change follows from the introduction of new technologies, is wrong. The eventual social and economic uses of a technology often turn out to be far removed from those originally envisioned. Users will do things with technology which the creators of that technology never dreamed off. We must accept this and work with it, inviting our users to join in the innovation process in a spirit of openness.

In Chapter 2, we continue the people-centric theme, but looking at people in an organisational context. We look at social networks and their impact on knowledge management. The chapter presents findings on what is the optimum size for a social network. It also reports on attempts to measure the gains achieved through the use of personal productivity tools such as email, further rebutting Solow's 'observation' quoted above. Looking to the future, the chapter sees future knowledge management systems building on current experience of Web 2.0, whilst the analysis of collaboration structures will partly formalise hitherto informal processes, thereby enabling the re-use and optimisation of knowledge processes.

Chapter 3 reminds us that users experience ICT through devices; understanding the way devices will evolve is fundamental to understanding how ICT will evolve. New modes of interaction will be introduced, and increased intelligence and awareness of context will generate easier to use devices. Whilst device evolution will lead us to an 'information anywhere, anytime' world, it will also raise questions of privacy and trust.

The final chapter in the first part of our book, Chapter 4, also discusses trust, but from a rather different standpoint. Specifically, it looks at how corporations can engender trust in their customers through customer advocacy; that is to say, through being on the side of the customer and providing the customer with all the information he or she needs to make purchasing decisions, including information about rival products. On the web site, this information can be provided in whatever form suits the customer's cultural background and style of decision making, where that style is learned from the customer's browsing behaviour.

Building the Infrastructure

We start this part by looking in Chapter 5 at the future of the web, which we believe will become a semantic web, with the existence of metadata (data about data) describing the meaning of web information. As a result, more intelligent applications will be able to be built; more intelligent search engines will be at our disposal, and richer information will be available. Meanwhile, the same technology will be used within organisations as a tool for systems and data interoperability and to unlock the knowledge within the unstructured information found in the organisation's reports, emails, intranet pages, etc.

Chapter 6 then describes how ICT infrastructure is becoming more flexible so as to respond to rapid changes in the business environment. The focus is on the physical infrastructure of computers and networks that underpin all ICT applications. An architectural approach based on loosely coupled services is described. This leads to the concept of a service-oriented infrastructure, where networks, servers, and storage hardware are made available as a set of services. Open standards are essential to achieving the full benefits of a service-oriented approach, and the chapter describes some of the work in this area.

A service-oriented infrastructure makes possible the flexible sharing of resources between organisations, where each organisation has access to virtualised resources. This brings new challenges in security, and Chapter 7 describes how those challenges can be met through secure business-to-business (B2B) gateways enacting security policies which are context- and content-dependent, and also fine-grained, i.e. relate to specific resources, not the whole of an organisation's resources.

Chapter 8 then looks at the physical network which all connectivity ultimately depends on. Future networks will be driven by a demand for richer broadband services with greater video content at increasing definition and quality. For the first time in the history of telecommunications, the same intrinsic technology is being used in the access and in the core networks. As a result, conventional architectures will not be able to scale to meet user demands in an economic way. The solution is an all-optical network which will massively reduce the number of electronic nodes in, and hence the cost of, the network.

Customers for ICT solutions require service level agreements (SLAs) to specify precisely what they can expect, and what recompense they can have when they do not get what they expect. Traditionally, such SLAs have been couched in the language of the service provider's business, not the customer's, and have described individual components of an ICT solution. The requirement today is for end-to-end SLAs describing the overall expected performance of the completed system, and related to the customer's applications, not the supplier's technology. How this is being done is the subject of Chapter 9.

Mobility is central to many of today's ICT systems, and Chapter 10 explores the links between mobility, authentication, identity, entitlement, and trust. It is argued that awareness of and competence in how and when to successfully manipulate these links is a critical aspect of successful ICT design. Trusted relationships will become a key differentiator for customer satisfaction associated with ICT based solutions and is of particular importance for ICT solutions involving mobility. After discussion of these issues, the chapter concludes with a longer-term look at the application of mobile ICT in virtual spaces.

Our next chapter, Chapter 11, talks about a technology whose impact is judged by how little we are aware of it. Pervasive computing is about computing 'woven into the fabric of life'. The chapter describes three broad strands. Pervasive information is about accessing any kind of information, via any kind of portal, from any information source, at any time, from anywhere. Pervasive sensing is the ability to acquire information about almost every object, person, or environment. Pervasive intelligence allows disperse networks of autonomous or semi-autonomous devices to gather data, reason about that data, and initiate required actions.

Chapter 12, the final chapter in the technology-focussed part of our book, looks at a very specific technology, that of artificial intelligence (AI). Wrongly seen by some as a failure in not living up to its initial promise, AI has given rise to genuine advances which are now part of mainstream computer science. After reviewing how the discipline has arrived at its current state, the chapter looks ahead to predict what AI might bring us in the future.

Applying Technology

Chapter 13 begins this part of our book by looking at how ICT is transforming the healthcare sector, using technology to cut across existing organisational and geographic boundaries. The chapter looks at a range of areas where ICT is having a significant impact, including telemedicine, telecare, the development of international healthcare standards for sharing health information, and the use of robotics in surgery and care.

Chapter 14 looks at ICT in the retail supply chain and suggests that any particular retailer can be located at a point on a spectrum between being totally efficiency-centric and totally customer-centric. Those at the former end of that spectrum will be primarily concerned with forecasting, promotion, and price-setting algorithms and assortment and allocation software. Those at the latter end will be more concerned with high resolution customer segmentation, use of 'loyalty' programs, managing high granularity customer buying patterns, and multi-channel retailing for an enhanced customer experience.

Chapter 15 looks at ICT in financial markets, specifically focusing on global financial markets which provide perhaps the most challenging customers for ICT vendors. The chapter reviews the impact of technology evolution on this market, arguing that technology is likely to revolutionise and reshape the companies that operate in the area.

In the last of our application-oriented chapters, Chapter 16, we look at a sector which is not, in the layman's eyes, normally associated with advanced technology, that of law. Detailed studies of lawyers' working practices reveal, in fact, that they do make significant use of technology. The chapter particularly discusses email management, a significant problem for lawyers, compounded by the onerous compliance requirements which they face. The chapter suggests that the use of semantic web technology is particularly relevant to supporting lawyers through a range of their activities.

Final Words

The final part of our book contains only two chapters. In the first of these, Chapter 17, we invite a futurologist to take a rather longer-term view than has been evident in the previous chapters. Rapidly covering a broad range of concerns, the chapter stresses the new experiences which technology will offer us and the need for the same technology to simplify the interfaces with those experiences. Humans will be displaced from more and more traditional roles; their real value will frequently be found in roles which require empathy with others.

In Chapter 18, we conclude our book by stepping back to identify the fundamental trends which underlie everything which has been described in our book. We find five. The first three are technological: the trend towards describing ICT as a set of services; the use of semantic technologies to enable interoperability and to exploit the richness of information within an organisation and on the web; the pervasiveness of these technologies, both in the literal physical sense and in the sense of their entering every aspect of our lives. The fourth trend is organisational: the increase in collaboration both between organisations and individuals. The final trend is purely social: the use of the net as a place for private individuals to meet and interact. We also discuss security, not a trend as such, but always a requirement when we deploy our technologies.

The User as Innovator

The intention of our book is to provide a broad overview of ICT: the component technologies, how they interact with users, and how they are used to create applications. We hope our readers will enjoy reading this book as much as we have enjoyed creating it. More than that, we hope our book will stimulate our readers to be innovative in how they themselves use these technologies. Both our first chapter and our final chapter observe that innovation is often created by the users of a technology, not just those who initially build the technology. Innovative new uses of technology and innovative adaptations of technology arise frequently from user communities. We hope our book will stimulate its readers to be innovative themselves in how they use and adapt ICT in their own professional and private lives.

References

Oulton, N. and Srinivasan, S. 2005. Productivity Growth and the Role of ICT in the United Kingdom: An Industry View, 1970–2000, CEP Discussion Paper No 681, Centre for Economic Performance, http://cep. lse.ac.uk/pubs/download/dp0681.pdf

Solow, Robert M. 1987. "We'd Better Watch Out," *New York Times Book Review.* July 12, 1987, p. 36.

Part One

People and Technology

1

Predicting the Socio-technical Future (and Other Myths)

Ben Anderson and Paul Stoneman

1.1 Introduction

Much human conduct is designed to avoid hazards and to promote beneficial returns. Indeed, this is the premise of the notion of 'risk societies' (Bcck and Beck-Gernsheim, 2002) where individuals rely on past and current information to determine their future, predominantly risk aversive, behaviours. This idiom of human affairs embraces most areas of life. Meteorologists can (sometimes) help us avoid bad weather; seismologists can warn of areas of pending earthquakes and volcanic eruptions; economists inform businesses and governments of forthcoming growth trends and market stability; political scientists tell us which party is most likely to form the next government, and climatologists warn of the dire consequences of global warming. Of course, the central tenet of forecasting and prediction is that by studying past information, we can – with some 'reasonable' degree of accuracy – project what is likely to happen in the future. However, there is no such thing as an exact science, and the confidence with which we can predict the future depends on the phenomena in question, the information available, and the granularity of prediction that we require.

This chapter concerns itself with predicting the future 'social implications' of information communication technologies (ICTs). It is sobering to remember that the telephone was not originally conceived as a means of human to human (or human to machine) communication. This form of usage evolved over time often in direct contradiction to notions of 'proper use' (Pool, 1983; Fischer, 1992). Despite this, it became the driving revenue stream for all telecommunications companies. Indeed, recent empirical studies of attempts at futurology have suggested that, amongst other problems, major reasons for failure have been an overemphasis on technology determinism, a poor understanding of social trends and change, and finally, the over-reliance on a

ICT Futures: Delivering Pervasive, Real-time and Secure Services
Edited by Paul Warren, John Davies and David Brown
Chapter 1 © 2008 University of Essex

linear progression model of change (Geels and Smit, 2000; Bouwman and Van Der Dun, 2007).

Here we examine long-term societal trends in behaviour using time-use data from the 1970s to the new millennium to show that in most measurable ways, the undoubted pervasiveness of modern information and communication technologies has had little discernable 'impact' on most human behaviours of sociological significance. We contrast this with observations from qualitative studies that illustrate how ICTs are changing the ways in which these behaviours are achieved, in other words, how ICTs are increasingly mediating (rather than impacting) everyday social practices. Historians of technology such as David E. Nye (Nye, 2006) remind us that human society co-evolves with the technology it invents, and that the eventual social and economic uses of a technology very often turn out to be far removed from those originally envisioned. This position enables us to think more clearly about the ways in which people's behaviours adapt to technologies and how supply and demand side interaction can lead to the co-adaptation of technologies.

1.2 Implicit Predictions

It is widely accepted that the natural world is governed by causal laws that provide a certain level of predictability for natural phenomena such as weather systems and animal behaviour (Hume, 1748; Hempel and Oppenheim, 1948). The social world, populated by slightly more anarchic humans, is less certain. This has led some critics to argue that attempts at predicting human behaviour may be fruitless (Hart *et al.*, 2007). However, despite huge problems predicting the future behaviour of individuals, it is possible to make reasonable predictions about *groups* of individuals (see, for example, Clark, 2003). A Humean view of society suggests that much like natural phenomena, human thought and behaviour are also governed by hidden laws creating stable and repeated outcomes. On this reading, human behaviour is far from random; we can, with the appropriate data, describe similar patterns of thought and behaviour across individuals and, combined with the appropriate methods, explain at least in part why they think and act the way they do.

Despite this, social scientists still see the primary role of theory as a way of providing explanations, not predictions (Popper, 1959). But the move from explanation to prediction is only one short hop – it could be argued that every explanation necessarily contains within it a prediction. Let us take a pertinent example. Climate change has focused much attention on the atmospheric movements of greenhouse gases, notably CO_2. An understanding of future emissions and movements helps us to predict the degree of hazard global warming (as a result of human activity) will pose on human populations.

> One way to gain perspective about the potential future trajectory for atmospheric CO_2 is to examine the geologic record of its concentration in the past. How high has the CO_2 concentration been in the past? How fast did it reach past high levels? (Shlesinger, 2003)

Modelling the relationship between previous levels and previous global temperatures provides an insight into how future levels of CO_2 emissions might impact global temperatures. Of course, the past and future impacts will never be the same – no two time periods will possess exactly the same conditions. But by making use of past information to make future projections, we move out of the realm of explaining only, and in terms of the future, out of the realm of guesswork and into the realm of forecasting and predicting. By doing so, futures become less uncertain.

1.3 Socio-technical Futures

The future of democratic societies is often couched in social-technical terms such as the creation of information societies (May, 2001) and e-democracies (OECD, 2004). Such futures often imply radical transformations perhaps by creating new forms of civic engagement or by helping social life to flourish. The main reason that such radical changes can be hypothesised is the huge dispersion of ICTs in society. Indeed, the most prominent of all ICTs, the Internet, can be thought of as the 'new television' in terms of uptake and usage. In 1950, only 10% of Americans had a television set; by 1959, this figure had soared to 90%, 'probably the fastest diffusion of a major technological innovation ever recorded' (Putnam, 1995). The Internet demonstrates similar figures. For example, in the UK, between 1999 and 2005, the number of people (as a percentage of the adult population) who went online rose from 14% to 61%.[1] In the US, 69% of the adult population now have internet access,[2] compared to only 20% 6 years ago.

Two schools of thought have emerged for possible internet effects – the utopians (Baym, 1997; Tarrow, 1999) and the dystopians (Nie and Hillygus, 2002). The first set of scholars believes that the Internet will restore a sense of community by providing a virtual meeting place for people with common interests (such as astronomy), which overcomes the limitations of space and time and where online communities could promote open, democratic discourse (Sproull and Kiesler, 1991), allow for multiple perspectives (Kapor, 1993), and even some political scientists believe it will help to mobilize collective action (Tarrow, 1999). Governments acknowledging the positive role of social capital and efficient information exchange have likewise posited the Internet as a radical and positive driving force for democratic society:

> Broadband enabled communication, in combination with convergence, will bring social as well as economic benefits. It will contribute to e-inclusion, cohesion and cultural diversity. It offers the potential to improve and simplify the life of all Europeans and to change the way people interact, not just at work, but also with friends, family, community, and institutions . . . (CEC, 2005)

Other scholars express reservations, providing two responses to the previous arguments. First, not all uses of the Internet are social – the predominant activities are the ones based around seeking information and engaging in solitary recreations (Nie and Erbring, 2000). Second, many 'social' activities online (for example, email) are asynchronous; responses and feedback are delayed until the recipient signs on, reads the

---

[1] http://www.statistics.gov.uk/CCI/nugget.asp?ID=8&Pos=6&ColRank=1&Rank=192.
[2] http://www.internetworldstats.com/stats2.htm.

message, decides to answer, and the original sender eventually receives the answer. If this is being done instead of a phone call or perhaps even a face-to-face meeting, then such virtual communication will give off the impression of maintaining the relationship, whilst in fact the quality of the relationship severely suffers (Gustein, 1999). This is a way of saying that the Internet may be diverting people from 'true' community relations.

The implications for internet-based socio-technical futures on either reading are clear; either it will continue to facilitate the growth of social and civic society or it will undermine them. Both of these arguments represent a type of thinking characterised as the snooker ball model.

1.4 The Snooker Ball Model

The causal nature of this model is essentially Newtonian and characterised by two conditions. First, there is a clear observable effect between a causal (independent) variable and an outcome (dependent) variable. For example, the presence or level of internet usage is an independent variable, and social and civic participation is the dependent variable. Second, that this relationship is unconditional, that is, wherever the presence of the causal variable is found, so too is the outcome variable (hence, a deterministic model). These clear and necessary impact effects are summarised in Figure 1.1.

Here we can see how the model derives its name. Patterns of behaviour are happily meandering along until individuals begin to adopt ICT usage and then are suddenly cannoned off into a new direction creating a different outcome. As long as the ICT usage prevails, the effect will be sustained creating a new equilibrium outcome.

Figure 1.2 demonstrates how in empirical terms we can observe the presence of such effects. Time runs along the X axis, whilst the vertical Y axis plots levels of a behavioural outcome, such as levels of social interactions. Between time points 1 and 3, social life exists without mass ICT usage. At time point 3, major ICT uptake begins so that by time point 9, a clear behavioural change can be observed. The three lines represent three different hypothetical scenarios. The lower dashed line represents a lasting negative effect of ICT usage, the middle straight line no effect at all, and the upper rising line a positive effect.

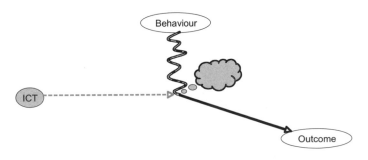

Figure 1.1   The snooker ball model

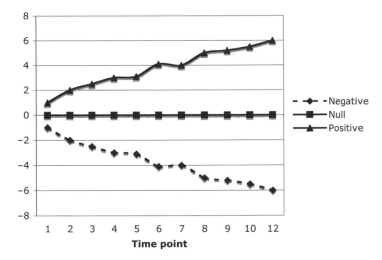

Figure 1.2   Observing the snooker ball model

As long as time series data are available on both the important variables – ICT usage and behavioural outcomes – then it is possible to test which of the three hypothetical scenarios actually apply to the effects of recent ICT uptake and usage. It turns out that historical time-use data that are now becoming available suggest very little significant change in the major uses of time over the last 20 years as uptake and usage of the Internet and mobile telephony has exploded in developed nations. Figure 1.3 below outlines data from time-use diaries that demonstrate that the average time in each of four developed nations allocated to employment, social activities, watching TV, and reading demonstrates little or no variation at all across time (see also Partridge 2007).

In other words, in most developed societies, we see that shifts towards 'e-societies' and the widespread ICT usage it brings does not translate into signs of significant social transformations despite the promulgations of various futurologists (Bell, 1973; Harvey, 1997; Castells, 2000).

The data suggest, therefore, that a Newtonian/deterministic view of technology is not just simple but simplistic. So, whilst Newtonian physics made way for quantum mechanics, the simple snooker ball model of socio-technical futures must likewise give way to a more nuanced understanding of cause and effect between society and technology.

1.5 The Conditional and Co-adaption Model

If the past is prologue, then the preceding 10 years suggest that not much will happen with socio-technical futures – at on level, the everyday lives of citizens look very much the same now as they did 30 years ago. But is that the end of the story? There are two reasons to think not. The first reason is similar to the weakness of Newtonian physics. Once mainstream science began to reduce the world to a lower level of analysis (particles and sub-particles), it was clear a new way of thinking was required. Cue quantum

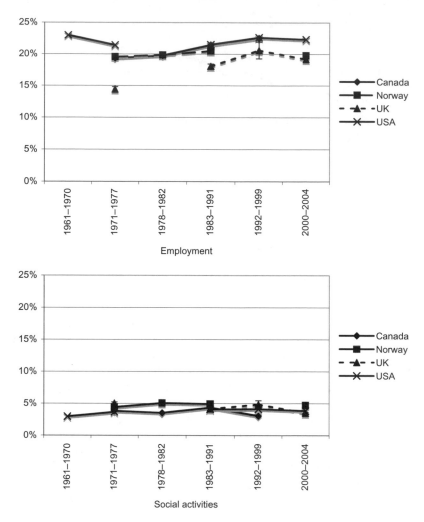

Figure 1.3 Historical trends in time use (Multinational Time-Use Survey, mean % of 24 hours for weekdays, all aged 20–59; error bars shown for UK only are ±95% confidence intervals)

mechanics. The second reason is an issue of time. Whilst we cannot currently observe widespread positive or negative ICT effects, this might be because users have not yet fully adapted to the potential usages of such tools.

Reducing socio-technical issues to a lower level of analysis requires a move away from whole population models and population averages. Specifically, this means rethinking the snooker ball model by abandoning the unconditional assumption described above. The world is just too messy for this to be the case. As soon as this condition is ditched, the idea of heterogeneity within a population and thus heterogeneity in responses to technological innovation becomes a primary empirical concern. Internet usage, for example, might facilitate further civic participation for some groups of people but not for others.

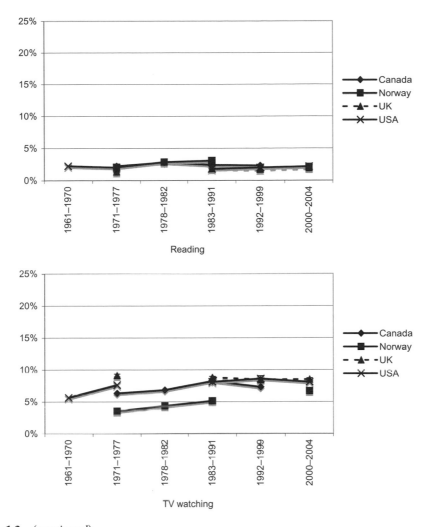

Figure 1.3 (*continued*)

As an example, Figures 1.4 and 1.5 present graphs of general election turnout by levels of internet usage (for political information) throughout the last British election campaign using the British Election Study 2005. In Figure 1.4, the sample population has been restricted to those individuals that express high levels of political interest and, as we can see, for these people, more or less internet usage for information seems to have no relationship with propensity to vote. The implication is that the strong level of political interest already found within the sample overrides any possible internet effects.

Figure 1.5 however, paints a different picture. The working sample for this analysis was restricted to those that express none or low levels of political interest. As the figure demonstrates, for this group, internet usage seems to boost turnout at higher levels.

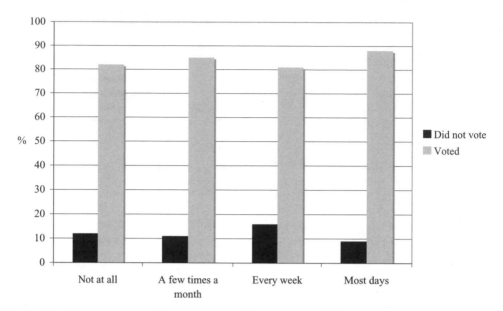

Figure 1.4   Vote turnout and internet usage (all persons with high levels of political interest)

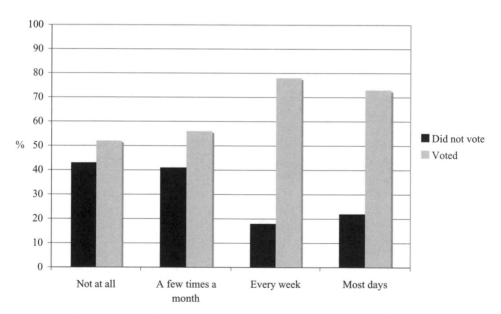

Figure 1.5   Vote turnout and internet usage (all those without or with little political interest)

So, by incorporating heterogeneity based on different levels of political interest into the analyses and by moving away from 'one model fits all' type of thinking, it is possible to demonstrate *conditional* positive internet effects.

Whilst belonging to a type of group can represent a conditional factor for internet effects, so too can the length of internet usage. The concern here is that relatively new users might not have had time to change their behaviour as a result of going online perhaps due to inexperience of the technology. Conversely, it might be the case that new users suffer from a 'new toy' effect so that the Internet is used more than will be the case. Both scenarios demonstrate the importance of allowing users to settle on a normal pattern of usage before any judgements are made regarding its impact on behaviour.

By making use of longitudinal time-use diary data from the Home Online survey (Anderson and Tracey, 2001; Anderson, 2005), it is possible to demonstrate this because we have 'before' and 'after' measures of behaviour (cf. Figure 1.2). Using multivariate regression analysis, it is possible to isolate whether more or less time spent on the Internet correlates with more or less time spent socialising as well as a range of other variables (Stoneman, 2006). The results show that in the first year of internet use, time spent surfing the Internet has no significant correlation with time spent socialising. However, further analyses demonstrate that for this data, time spent web browsing is mainly substituting time spent watching TV and doing nothing.

Repeating the analysis for a longer time frame (the first 2 years of internet use), on the other hand, demonstrates small but significant and negative internet effects on time spent socialising. Although the effect of internet use on socialising is marginal (every extra hour spent online produces a net effect of 6 minutes less on social activities), the difference in the two sets of results demonstrates the importance of allowing users to settle upon a normal routine of usage before searching for possible effects.

The implication of these analyses is that, far from having a direct impact on populations, technologies conditionally co-adapt with social life. Across time, individuals may or may not use innovations such as ICTs to support current behaviours and might even occasionally, once familiarity with the tool is established, create new patterns of behaviour. Socio-technical change is thus far from simple, and as a corrective to the snooker ball model, we must turn to a co-adaptation model.

In this model, we see that behaviours and usages are not straightforwardly predictable from the affordances of the ICTs as they depend on a range of contingent and contextual factors including life stage, skills, needs and resources. Users may adapt their behaviour to make use of the affordances of the ICTs and in turn, the ICT producers act on these new behaviours. Some behaviour may simply be a continuation of the past, whilst others may be generally new or disruptive (Gower *et al.*, 2001) in an ongoing process of domestication (Silverstone and Hirsch, 1992; Haddon, 2006).

1.6 Feedback Mechanisms and the Evolutionary Model

The line of thinking behind the co-adaption model is now well established in social scientific studies of technology (Bijker *et al.*, 1987; MacKenzie, 1998; Nye, 2006). By positing conditional effects, we begin to speak of *probable* causes and outcomes as

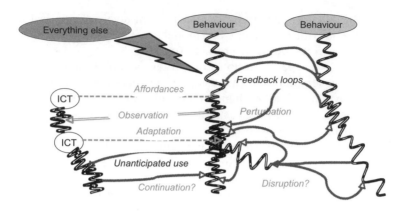

Figure 1.6   The evolutionary system model

opposed to the rather crude and simple 'if A then B' logic, and we also begin to speak of iterative cycles of causal processes. But how far does the uncertainty principle go in the realm of socio-technical futures? If we observe a probable cause of an outcome now, how sure can we be that this will occur again in the future?

A major problem is that the interactions between people, technologies, and the producers of those technologies are rather more complex even than this. For one thing, technologies have always generated unanticipated uses. This tradition of finding uses for technologies that their designers and marketers did not perceive still continues. This issue has been well known in the field of computer supported cooperative work (CSCW) for some years (Robinson, 1993) where it is common place for users of office systems to adapt them to their own purposes in ways that have not been foreseen by the designers of the systems. Rather than viewing these new uses as 'improper' or 'user error', CSCW sees in them an opportunity to capture users' creativity and to fold these uses back into the product or system. In other words, the users become the co-designers of the system, and the result is usually a workplace system that is far better suited to the work practices of those users and is therefore far more likely to be used.

We attempt to capture this complexity and reflexivity in terms of an evolutionary co-adaptation model that is intentionally represented as complex and dynamic (Figure 1.6). Here we can see the myriad of feedback loops that exists between behaviours directly and indirectly related to ICT usages, and in turn the complex and iterative evolutionary relationship between these and the design and production of ICTs. In this model, the term evolutionary comes to the fore as we suggest that usages (behaviours) and technologies are engaged in a form of evolutionary co-adaptation where changes on one 'side' are intimately related to changes on the other and can frequently lead to startling innovations (Krebs and Davies, 1997).

In the context of predicting future technology usages, this model inevitably reduces our confidence in our ability to make sensible statements about what might happen, and it is, in particular, the likelihood that unanticipated usages (and thus revenues) will come to the fore that drives this uncertainty. Who would have predicted that SMS (short message service) would turn out to be used as a de facto means to keep in touch, to gift friendship, to dump boyfriends, to send intimate images, to organise protest, or to look busy (Rheingold, 2002; Ling, 2004)? Or that Bluetooth messaging would be used to gift pornography, to send threats, to project sexualities, and to hide identities (Bond, 2007)?

It seems plausible, therefore, that the diversity of users and usage contexts for *malleable* consumer technologies means that both of these may be unknowable at the time of conception and design. Thus, perhaps the most fundamental challenge facing the design of products and services is that what a thing is for, and who its users will be, can rarely be defined in advance. This means that any approach to the design of artefacts that assumes that the confident definition of user, task, and goal is possible will be of limited use (Lacohee and Anderson, 2001).

1.7 Implications for Forward Thinking

The crucial aspect captured by an evolutionary understanding of socio-technical futures is the importance of a feedback mechanism between innovators and users. The implication is that innovators perhaps need to innovate less and to listen and observe more whilst placing more control over product development in the hands of customers. Whilst most commercial organisations would see this as deeply threatening, new models of customer-led innovation and indeed customer-generated innovation are showing how these phenomena can be turned to profit. Such models can provide essentially free research and development (the customers do it themselves), will con struct services 'developers' would never have dreamed of, will meet customers' own heterogeneous needs, and can generate unanticipated revenues.

These models acknowledge that many people like to consume; many like to customise, and more than are commonly supposed will seek to construct products and services for themselves (Oudshoorn and Pinch, 2003). Drawing on a range of empirical studies, Eric von Hippel, for example, estimates that up to 40% of users actively develop or modify products (von Hippel, 1986). First mooted under the rubric of mass customisation (Davis, 1987) and more recently observed and advocated again by von Hippel (2005) amongst others, the model involves the potential users/customers in a rapid cycle of design, use, and redesign – betaworld.

To give one excellent example, a recent study of the creation of a teen-oriented web site recorded the rapid transition of the site from an 'Editors know best: we create, you use' model to a 'They know best, we supply the framework, they do the content model' (Neff and Stark, 2004). Indeed the executive interviewed said:

> We don't have people sitting around thinking. 'What do teens want?' It doesn't work. Even if you could figure it out, it wouldn't last. You can try to write for them but it doesn't work. Now 95% of our content is written by teens themselves. (Neff and Stark, 2004)

This model can now be seen at work in a plethora of internet-based services that implicitly or explicitly support users in the creation of their own content (blogs, wikis, flickr, facebook, youtube, etc.) and their own applications and services (google, facebook, ning). It is a model radically different from the traditional 'innovate, design, build, launch, market, sell, wait for revenue' pipeline model because it allows the business case/model to evolve during development, and thus to respond to unanticipated use, rather than being (usually) incorrectly specified in advance. It may be that such *adaptive revenue models* are one part of a response to the problem of responding to disruptive technologies. If uses and thus revenues cannot be predicted in advance, then at least we can put in place adaptive organisational mechanisms so that emerging uses and revenues can be exploited rapidly. This requires commercial organisations to admit both institutionally and emotionally that they are no longer in control of their product and service lines.

This, then, is our partial answer to the problem posed in the introduction. Rather than persisting in attempting to predict future ICT usage and revenue models, and thus producing future visions and business cases that turn out to be wrong, we suggest that truly participatory designs (Bjerknes *et al.*, 1987), grounded innovation (Anderson *et al.*, 2002), open systems, and adaptive revenue models can lead us to a more effective, flexible, and responsive innovation process. On current trajectories, it will certainly provide us with an explosion in novel tools with which to live our lives. In terms of who uses what for which purpose, this does not make socio-technical futures more certain. It does, however, acknowledge that socio-technical futures now passed were *never* certain, and there is no reason to believe current futures will be any different. Nonetheless, our answer is not simply que sera sera. By working with uncertainty as opposed to projecting regardless of it, one thing will become more certain – a more user-informed innovation process leading to significantly improved business models.

Acknowledgements

This research was supported by the UK Economic and Social Research Council (ESRC) funded e-Society Programme project 'Using Time-Use Data to Analyse Macro and Microsocial Change in an e-Society' (RES-341-25-0004) – http://www.essex.ac.uk/chimera/projects/esoctu/.

Earlier forms of these arguments were presented at an ESRC e-Society Workshop in 2006 entitled 'The Internet in Britain: Statistical Analyses' and at a BT Masterclass in 2007 entitled 'Social Innovation in a World of Malleable Technology'.

References

Anderson, B. (2005) The value of mixed-method longitudinal panel studies in ICT research: Transitions in and out of 'ICT poverty' as a case in point. *Information Communication & Society*, 8, 343–67.
Anderson, B. and Tracey, K. (2001) Digital Living: The 'Impact' or otherwise of the Internet on Everyday Life. *American Behavioral Scientist*, 45, 457–76.
Anderson, B., Gale, C., Gower, A. P., France, E. F., Jones, M. L., Lacohee, H. V., McWilliam, A., Tracey, K. and Trimby, M. (2002) Digital living – people-centred innovation and strategy. *BT Technology Journal*, 20, 11–29.
Baym, N. (1997) Identity, body, and community in on-line life. *Journal of Communication*, 47, 142–48.
Beck, U. and Beck-Gernsheim, E. (2002) *Individualization: Institutionalized Individualism and Its Social and Political Consequences*, London, Sage.
Bell, D. (1973) *The Coming of Post-Industrial Society*, Harmondsworth, Penguin.
Bijker, W. E., Hughes, T. P. and Pinch, T. J. (1987) *The Social Construction of Technological Systems: New Directions in the Sociology and History of Technology*, Cambridge, MA; London, MIT Press.
Bjerknes, G., Ehn, P., Kyng, M. and Nygaard, K. (1987) *Computers and Democracy: A Scandinavian Challenge*, Aldershot, Avebury.

Bond, E. (2007) mobile phone = bike shed? Children, mobile phones and sex. *Chimera Seminar Series.* Chimera, University of Essex.

British Election Survey 2005. See http://www.essex.ac.uk/bes/2005/index2005.html

Bouwman, H. and Van der Dun, P. (2007) Futures research, communication and the use of information and communication technology in households in 2010: A reassessment. *New Media & Society*, 9, 379–99.

Castells, M. (2000) *The Rise of The Network Society, The Information Age – Economy Society and Culture, Vol 1.*, London, Blackwell.

CEC (2005) Working together for growth and jobs: A new start for the Lisbon Strategy. *Communication to the Spring European Council.* Brussels, Commission of the European Communities.

Clark, A. (2003) Self-perceived attractiveness and masculinization predict women's sociosexuality. *Evolution and Human Behavior*, 25, 113–24.

Davis, S. (1987) *Future Perfect*, London, Addison Wesley.

Fischer, C. (1992) *America Calling: A Social History of the Telephone to 1940*, Berkeley, CA, University of California.

Geels, F. W. and Smit, W. A. (2000) Failed technology futures: pitfalls and lessons from a historical survey. *Futures*, 32, 867–85.

Gower, A. P., Lacohee, H., Jones, M. L., Tracey, K., Trimby, M. and Lewis, A. (2001) Designing for disruption. *BT Technology Journal*, 19, 52–9.

Gustein, D. (1999) *E.con: How the Internet Undermines Democracy*, Toronto, Stoddart.

Haddon, L. (2006) The contribution of domestication research to in-home computing and media consumption. *The Information Society*, 22, 195–203.

Hart, S. D., Michie, C. and Cooke, D. J. (2007) Precision of actuarial risk assessment instruments: Evaluating the 'margins of error' of group v. individual predictions of violence. *British Journal of Psychiatry*, 190, s60–5.

Harvey, D. (1997) *The Condition of Postmodernity: An Enquiry into the Origins of Cultural Change*, Oxford, Basil Blackwell.

Hempel, C. G. and Oppenheim, P. (1948) Studies in the logic of explanation. *Philosophy of Science*, 15, 135–75.

von Hippel, E. (1986) *The Sources of Innovation*, Boston, MA, MIT Press.

von Hippel, E. (2005) *Democratizing Innovation*, Boston, MA, MIT Press.

Hume, D. (1748) *Enquiry concerning Human Understanding. Sections IV–VII (paras. 20–61)*, Oxford, Oxford University Press.

Kapor, M. (1993) Where is the digital highway really heading? *Wired*, 94 (July/August), 53–9.

Krebs, J. and Davies, N. (1997) *Behavioural Ecology: An Evolutionary Approach*, Oxford, Blackwell Science.

Lacohee, H. and Anderson, B. (2001) Interacting with the telephone. *International Journal of Human Computer Studies*, 54, 665–700.

Ling, R. (2004) *The Mobile Connection: The Cell Phone's Impact on Society*, London, Morgan Kaufmann.

Mackenzie, D. (1998) *Knowing Machines: Essays on Technical Change*, Boston, MA, MIT Press.

May, C. (2001) *The Information Society: A Sceptical View*, London, Polity Press.

Neff, G. and Stark, D. (2004) Permanently beta: Responsive organization in the internet era. In Howard, P. N. and Jones, S. (Eds.) *Society Online*, pp. 173–188. London, Sage.

Nie, N. and Erbring, L. (2000) Internet and mass media: A preliminary report. *IT&Society*, 1, 134–141.

Nie, N. H. and Hillygus, D. S. (2002) The impact of internet use on sociability: Time-diary findings. *IT & Society*, 1, 1–20.

Nye, D. E. (2006) *Technology Matters: Questions to Live With*, Cambridge, MA, MIT Press.

OECD (2004) *Promise and Problems of E-Democracy: Challenges of Online Citizen Engagement*. Paris, Organisation for Economic Cooperation and Development.

Oudshoorn, N. and Pinch, T. J. (2003) *How Users Matter: The Co-construction of Users and Technologies*, Cambridge, MA; London, MIT Press.

Partridge, C. (2007) Changing patterns of time use in an e-society: A historical and comparative descriptive analysis. *Chimera Working Paper 2007-06*. Ipswich, Chimera, University of Essex.

Pool, I. D. S. (1983) *Forecasting the Telephone. A Retrospective Technology Assessment*, Norwood, NJ, Ablex.

Popper, K. (1959) *The Logic of Scientific Discovery*, London, Hutchinson.

Putnam, R. (1995) Bowling alone: America's declining social capital. *Journal of Democracy*, 6, 65–78.

Rheingold, H. (2002) *Smart Mobs*, Cambridge, MA, Persius.

Robinson, M. (1993) Design for unanticipated use. In De Michelis, G., Simone, C. and Schmidt, K. (Eds.) *ECSCW '93: 3rd European Conference on Computer-supported Cooperative Work*. Milan, Dordrecht, London, Kluwer Academic Publishers.

Shlesinger, W. H. (2003) The carbon cycle: Human perturbations and potential managements options. In Griffin, J. M. (Ed.) *Global Climate Change: The Science, Economics and Politics*. Cheltenham, Edward Elgar Publishing.

Silverstone, R. and Hirsch, E. (1992) *Consuming Technologies: Media and Information in Domestic Spaces*, London, Routledge.

Sproull, L. and Kiesler, S. (1991) *Connections: New Ways of Working in the Networked Organization*, Cambridge, MA, MIT Press.

Stoneman, P. (2006) Exploring time use – a methodological response to 'web-use and net-nerds'. *Chimera Working Paper 2006–11*. Ipswich, Chimera, University of Essex.

Tarrow, S. (1999) *Power in Movement. Social Movements And Contentious Politics*, Cambridge, MA, Cambridge University Press.

2

Social Networks, Social Computing and Knowledge Management

Nicholas J. Kings, John Davies, David Verrill, Sinan Aral,
Erik Brynjolfsson and Marshall van Alstyne

2.1 Introduction

Social networks are the 'connections' that individuals utilise through various modes of communication to conduct their work. These communications come in the form of email, instant messaging, phone calls, wikis, blogs, face-to-face interactions, etc. These connections make up the central nervous system of information intensive organisations. The information that workers use to analyse, explore, understand and make decisions about their environment often flows through these connections. Understanding these connections can therefore provide important insights into how the information structure of a firm affects it performance.

Knowledge management is defined as the 'systematic application of actions to ensure that an organisation obtains greatest benefit from the information that is available to it' (Marwick 2001). Knowledge here is seen as the experience of the people within an organisation (tacit knowledge) combined with information from documents within the organisation, as well as access to relevant reports from the outside world (explicit knowledge). As such, knowledge management combines the use of ICT systems with an insight into the social processes around the movement and production of information.

Knowledge management systems are now moving to a position of exploiting the insights gained from analysis of social structure, and from the dramatic rise in popularity of social networking systems both inside and outside the enterprise. For example,

ICT Futures: Delivering Pervasive, Real-time and Secure Services
Edited by Paul Warren, John Davies and David Brown
© 2008 John Wiley & Sons, Ltd

the Facebook[1] system was seeing monthly rises in the number of unique visitors of 80% in mid-2007. Similarly, a recent survey revealed widespread interest in corporate use of social computing and related technologies (McKinsey 2007).

Considering knowledge management trends in recent years, we can distinguish three broad phases: first, the repository-centric view characterised by one or more central information repositories with a set of corporate contributors and reviewers; second, the move to smaller, facilitated knowledge communities; and finally, previous knowledge management archetypes have been broadened and complemented by a trend termed 'social computing' (Charron et al. 2006). Brown (2007) further describes the current situation:

> Front-line business people continue to tug at the fringes of hierarchical, bureaucratic, IT architectures by downloading and installing themselves productivity tools like desktop search, wikis and weblogs . . . the water cooler conversation, now lives somewhere between your instant messaging network, email discussion threads, and the public blogosphere.

The structure of this chapter is as follows. We begin with a discussion of the importance of analysing social networks with the enterprise. We then describe applications of social computing to knowledge management and conclude with a discussion of trends for future enterprise-wide social computing applications.

2.2 Do Social Networks Matter?

The answer to this question is simple: social networks matter far more than we have ever imagined.

In the USA, information workers account for nearly 70% of the labour force, and contribute more than 60% of the total value added to the US economy (Apte and Nath 2004). Information workers make decisions and actions that generate value-added information products, instantiated as advice, reports, designs or legal contracts that are subsequently sold at a premium.

If information is the critical input to 70% of the work being conducted in today's economy, then the social avenues and channels that distribute information among individuals, groups and populations are possibly the supply chains of knowledge-intensive industries. The networks of relationships that connect people are perhaps the most important avenues through which information flows in the business world, and social networks offer a structure for the systematic analysis and measurement of information flows in business networks.

Social networks are conduits for communication flows, and communication flows contain the lifeblood of companies: information. Information, and the knowledge instantiated within it, is a critical success factor in business. What if social networks could help answer questions such as:

- Does the size or shape of a person's social network matter to performance?
- Does a larger social network cause information overload and poor performance?

---

[1] http://www.facebook.com/.

- What are the communication and information practices of effective workers?
- How do ICT-enabled processes change work organisation and business processes in the context of information work?
- How do new ICT-enabled processes change communication flows and how do these flows in turn impact productivity and performance?

Past and recent research at the Massachusetts Institute of Technology (MIT) Center for Digital Business and at other leading universities, suggests that the answers to these questions are at hand. By collecting email, instant messaging, phone and face-to-face communications data, social networks reveal patterns of performance. These data offer a view into how information moves between information workers as they do their jobs. With this information in hand, individuals, managers and organisations have an opportunity to optimise workflow and to improve performance.

2.2.1 Social Networks and Measuring Individual Information Worker Productivity

While social networks can be used to uncover patterns of information flow that predict efficiency, quality, error rates or more repeat business, the focus of this section is productivity and the reasons why social networks matter to the productivity of information workers.

Ichniowski, Shaw and Prennushi (1997) and others were able to specify and measure production functions for blue collar workers, such as those working in steel finishing lines. This has greatly advanced the understanding of the effects of work practices and technology on productivity for these workers. In contrast, there has been little work measuring the productivity of information workers. Even measuring the impact of technology investment on productivity at the company level has proved difficult (Brynjolfsson and Hitt 1996). Since then, recent research has focused on explaining variance in the benefits gained through IT spend (Aral and Weill 2007), and modelling and measuring how information flows and ICT use impact the productivity of information workers.

Van Alstyne *et al.* (Aral *et al.* 2006; Bulkley and van Alstyne 2006; Aral and van Alstyne 2007; Aral *et al.* 2007) collected data from detailed access to three US-based executive recruiting firms and set out to analyse the data for patterns of information flow as indicators of productivity. Included in the data was information on 1300 projects, 125,000 email messages, accounting information, semi-structured interviews, a detailed survey and the regional socio-economic conditions in which the firm's services were delivered. Several measurable outputs were calculated from these sources, including:

- revenues per person and per project;
- number of completed projects;
- duration of projects;
- number of simultaneous projects;
- compensation per person.

The results obtained from analyses of these data matched with individual and collective social networks were profound. While the analysis is ongoing, the findings to date include a set of results on topics ranging from information overload to multitasking, project duration and completion, to incentive theory and sharing, to ICT skills, to social network structure and information advantage, and to information diffusion dynamics.

Here are three concrete examples:

1. Information Diffusion. Using the data gathered, Aral *et al.* (2006, 2007) explored a set of questions around the diffusion of information within a company, the impact of social networks and the use of information technology upon behaviours. What are the productivity effects of information diffusion in networks? What predicts the likelihood of receiving information diffusing through a network, and receiving it sooner? Do different types of information diffuse differently? Do different characteristics of network structure, relationships and individuals affect access to different kinds of information?

   The results indicate that those people who act as 'information hubs' in the social network are more productive; information hubs are those that communicate with a broad set of people within their own organisation and that have high communication volume. Aral and van Alstyne (2007) found that larger social networks deliver diverse information but with a decreasing rate of return. In turn, productivity increases with novel information, but again at a declining rate. Thus, there will be an optimum size for a person's communication network.

   In the case of these particular firms, adding more than 15 or 20 people to one's own internal social network had little additional impact on productivity, output or performance. In a different industry, the size of an optimum network may, for example, be 75 people: the optimum network size for a particular enterprise is dependent upon a number of variables, such as the type of knowledge work and the complexity of interactions.

   So, the recent media interest about 'aggressive networking' and the value of becoming a 'social butterfly' has real limits especially if we consider the time and effort required to make and to maintain relationships. The main insight highlighted the cognitive bottleneck of the human mind: while information is no longer scarce, attention is. The data indicate that employees with diverse social networks receive more novel information and are therefore more productive; productive people are those that connect to multiple different parts of an organisation. In essence, networks that deliver the right information are highly valuable, but adding more contacts is less valuable the larger your network becomes.

   So how is it possible to translate this analysis to the bottom line? Given the availability of team, project, accounting and compensation data, the research team was able to create a fine-grained analysis. For example, having access to new information sooner is a predictor of higher productivity. In the case of the recruiting firms studied, seeing an additional name of a new candidate within 1 week of its first emergence in the communication network was worth $321 to a recruiter's bottom line revenue generation, compared to $115 if seen within the first month. Timeliness and relevance were key variables in performing a search within the recruiting firms. For example, if a new candidate's name circulated quickly through the communication network, then that candidate could be included in ongoing recruitments.

2. Information, Technology and Multitasking. Popular press has suggested that new information and communication technologies are driving business at 'the speed of thought', or putting today's labour force on the 'information superhighway'. Initial estimates from the data analysis demonstrated strongly that the use of ICT was associated with more revenue generation and more projects completed per unit time, but that using ICT was also associated with longer average project duration. ICT was slowing work down but was also enabling workers to do more and to generate more revenue per unit time.

   ICT was changing how people worked, rather than just speeding up traditional ways of working. Recruiters who used the information systems of the firm more heavily and who were in positions to be information hubs in the social network structure of the firm were multitasking more and multitasking more effectively. ICT and access to information were enabling these information workers to handle more simultaneous projects without slowing down any one individual project completion rate. Up to a certain point, more multitasking improved productivity, but after reaching an optimum level, taking on additional work made employees less productive.

3. Information Overload. The pervasiveness of email has given rise to what many workers consider information overload. On a statistically significant basis, four key email traits suggest effective use of email. First, those people who are central nodes of information, who bridge structural holes in the social network, are significantly more effective workers. Second, those who send short email messages have higher productivity. Third, those who communicate via email with accurate, concise subject descriptions receive better email response rates. Fourth, those who invest in building their social network early in their careers are better able to exploit the network for productivity gains later.

The data from this research have identified a number of principles of effectively managing information in a social network with direct bottom line implications. Each of these principles has a basis in the underlying analysis of the social networks in three recruiting firms, but they may differ for other types of information workers:

- ICT does not just speed up tasks; ICT changes how those tasks are achieved.
- The productivity of an information worker is affected by how quickly his or her colleagues respond to information requests.
- Information workers who use the company's knowledge systems are more productive.
- Information workers who use ICT for higher value information processing are more productive.
- Information workers who gather information from a more diversified social network are more productive.
- Information workers who optimise the management of information contacts over time gain greater benefit over their career life cycle.

2.3 Knowledge Management Applications

The purpose of knowledge management can be broadly defined as seeking to maximise the benefit to an organisation of its information assets. Given the productivity benefits for information workers identified above, it is instructive to consider the place of social computing and the social networking capabilities it offers in the wider knowledge management context. Marwick (2001) suggests that effective knowledge management requires a combination of organisational, social and managerial initiatives alongside the development of appropriate knowledge access technologies. Furthermore, knowledge sharing software supports the activities to collate, categorise and distribute information (Davies *et al.* 2005).

It is important to state that there are different forms of knowledge that are processed differently by individuals and by the enterprise as a whole. In particular, it is useful to differentiate between explicit and tacit knowledge: Nonaka (cited by Marwick 2001) formulated a theory of organisational learning that focused on the transformation of tacit into explicit knowledge and vice versa. Tacit knowledge is held internally by each knowledge worker, and is formed by past experiences, beliefs and values. Explicit knowledge is represented by some document, such as a web page or a video, which has been created with the goal of relaying some piece of tacit knowledge from one person to another. Organisational learning occurs as people participate in shared activities, and their knowledge is articulated, making it available to others. Typical activities involved in converting from one form of knowledge to another are shown in Figure 2.1.

These transformational processes are happening all the time, as an individual moves between different work situations and communities (Wenger 1999; Marwick 2001; Shneiderman 2003). All of the processes are important, but the emphasis of one process over another is a matter of balance and choice in which knowledge management is deployed. Brown and Duguid (1991, 2000) suggest that learning and knowledge sharing is a key activity supported by a community of practice: learning is essentially a social activity. The importance of a document or the relevance of a piece of information, to a particular community, is a by-product of how that information has been propagated: the importance and meaning of information is negotiated rather than being fixed and agreed in advance.

Tacit to Tacit	Tacit to Explicit
Socialisation	*Externalisation*
Team Meetings, Discussions	Answering explicit questions, writing of a report
Explicit to Tacit	**Explicit to Explicit**
Internalisation	*Combination*
Reading and learning from a web page	Using a search engine, emailing a report to a colleague

Figure 2.1   Converting between forms of knowledge (after Marwick 2001)

Rogers (1995) suggests that people gather and share information unequally through their social networks, as a person tends to trust information from similar people. However, those people with strong social connections tend to only know similar information. Granovetter (1973, 1983) terms this as the 'strength of weak ties', where new or useful information tends to be found at a lower personal cost through weak social contacts. Constant *et al.* (1996) suggest that in a situation where opinions are required rather than a definitive answer, the more possible solutions, the better. Thus, community systems that allow access to a greater number of weak ties will have a better perceived utility. However, this has to be balanced against the cognitive load to understand the alternative suggestions.

Information gathering and information sharing are separate but connected activities (Hyams and Sellen 2003a, 2003b); information gathering involves a significant amount of collation and learning rather than directly finding a particular fact. Annotations that users will make for their own private usage will tend to be different from those annotations made for 'public' consumption (Marshall and Brush 2004). Information sharing requires that a feedback path is available from the information recipient back to the sender. Feedback from recipients is a powerful motivational tool to encourage sharing, but also allows information publishers to tailor information to more of the recipients' needs and business context (Golder and Huberman 2006).

Knowledge management can thus be seen, in part, as the use of social computing within the enterprise. Collaborative systems, such as the content creation process found in wikis, will enable communities within the company to reach a consensus via collaborative working. Social computing applications explicitly make use of collaboration, communities and network effects, where the utility of an application increases with a greater number of users (Hinchcliffe 2006; Heath *et al.* 2007): word of mouth recommendations greatly accelerate the take-up of such systems.

While, Wikipedia[2] has become one of the most referenced knowledge sources on the Internet, social computing technologies are not solely applicable on the Internet, but are also seen as viable enterprise solutions. Companies such as Dresdner Kleinwort Wasserstein are using wiki technologies to enable the co-creation of information, avoiding complicated email exchanges; wikis are seen as 'more participative and non-threatening' than other collaborative technologies (Nairn 2006). The lessons learnt from the growth of applications in the consumer space are being applied to enabling new ways of working and boosting employee productivity. Lightweight social computing technologies encourage sharing, as each person is able to easily contribute a small piece to a growing body of information.

Though ICT systems are not able to enforce an attitude to share information within an enterprise, facilities can be put into place to encourage and support those knowledge sharing activities (Kings *et al.* 2003, 2007; Davies *et al.* 2005; Bontcheva *et al.* 2006). In effect, the ICT system becomes a 'place' that enables or disables certain forms of interactions (Raybourn *et al.* 2003): a culture of sharing can be fostered by an awareness of the underlying social norms, and the artefacts within a system that expose support that culture.

---

[2] http://www.wikipedia.org/.

2.4 Future Research and Trends

From the perspective of the future of ICT, whether ICT delivers productivity benefits comes down to how technologies are used by individuals and organisations. This point goes a long way towards explaining why companies with the same ICT expenditure can perform so differently (Aral and Weill 2007). As with most research, several important new questions have emerged from our analyses. For instance, applying the same techniques to different industries and different types of workers may lead to different principles. Future questions to explore include the following:

- What technologies and practices are most effective for various types of work?
- How does collaboration correlate with performance, at the individual and organisational level? How does collaboration differ across different media (for example, IM compared to email)?
- What practices, incentives and culture are correlated with increased collaboration? What should managers do, and what should they avoid doing?
- How can qualitative measures improve our productivity measurement?

Future knowledge management systems will build upon the positive experiences of massively collaborative Web 2.0-type knowledge articulation processes, such as the tagging process (as seen, for example, in delicious[3] or flickr[4]), the collaborative content creation process found in wikis and the content sharing approach found in blogging systems. Currently, the power of these systems stems from their massively collaborative nature and their easy and intuitive handling while being restricted to resource annotation (in the case of tagging) and generation of interlinked document-style content (in the case of wikis and blogs).

Collaborative content annotation and creation process can be raised to the level of knowledge articulation with increased expressivity by exploiting semantic technologies, while still retaining intuitive and collaborative Web 2.0-style tools. Semantic technology offers a unified, more formal structure against which to annotate knowledge, facilitating shared access and more sophisticated information analysis, for example, by creating a task-specific search tool (Duke *et al.* 2007).

Mika (2005) suggests that the Semantic Web has been defined to facilitate machine understanding of the World Wide Web; however, the process of creating and maintaining that meaning is a purely social activity. Each ontology is created in a process that requires a group, or community, to build and share an agreed understanding of the important concepts for that community. An understanding of social presence is crucial in understanding how an ontology evolves and gains acceptance.

Much of the life of an enterprise is represented through the interaction of informal knowledge processes. Much of these interactions occur in the form of communications like emails, blogs, wikis, reports and other types of documents that are created, processed and archived within or in relation to the enterprise. Part of such document flows is regulated by business processes or policies, while the other part originates from

---

[3] http://del.icio.us/.
[4] http://www.flickr.com/.

semi-formal or informal activities of knowledge workers or communities within an enterprise. By analysing these collaboration structures with learning and mining methods, it should be feasible to extract the structure of informal processes and to formalise them to the needed degree. This will enable process reuse and optimisation on the personal and on the enterprise level (while maintaining privacy and confidentiality).

While processing the informal knowledge, a knowledge worker is frequently dealing with many tasks and much incoming information in parallel, and has to switch often between tasks and between various kinds of documents. Thus, a critical aspect of future knowledge systems will be the modelling of the user's differing contexts and offering support for context management. Here, context is comprised of various aspects including the user's current tasks, interest profile and experience, running applications, device, role, and so on. At times, the system will suggest to the user to switch context, and at others, to suppress incoming alerts and tasks which could lead to unwanted context interruption. Thus, the management of context, combined with formal and informal knowledge processes, will be key constituent of future knowledge management systems.

References

Apte, U. M. and Nath, H. K. (2004) Size, Structure and Growth of the US Information Economy. Available from: http://www.anderson.ucla.edu/documents/areas/ctr/bit/ApteNath.pdf [Accessed 16 October 2007].

Aral, S. and van Alstyne, M. (2007) Network Structure & Information Advantage. *Academy of Management Conference*. Philadelphia, PA. Available from: http://ssrn.com/abstract=958158 [Accessed 16 October 2007].

Aral, S. and Weill, P. (2007) IT Assets, Organizational Capabilities & Firm Performance: How Resource Allocations and Organizational Differences Explain Performance Variation. *Organization Science*, 18, 5, 1–18.

Aral, S., Brynjolfsson, E. and van Alstyne, M. (2006) Information, Technology and Information Worker Productivity: Task Level Evidence. *27th Annual International Conference on Information Systems*. Milwaukee, Wisconsin. Available from: http://ssrn.com/abstract=942310 [Accessed 16 October 2007].

Aral, S., Brynjolfsson, E. and van Alstyne, M. (2007) Productivity Effects of Information Diffusion in Email Networks. *28th Annual International Conference on Information Systems*. Montreal, CA. Available from: http://ssrn.com/abstract=987499 [Accessed 16 October 2007].

Bontcheva, K., Davies, J., Duke, A., Glover, T., Kings, N. and Thurlow, I. (2006) Semantic information access. In Davies, J., Studer, R. and Warren, P. (Eds.) *Semantic Web Technologies: Trends and Research in Ontology-based Systems*. Chichester, England, John Wiley & Sons, Ltd.

Brown, M. (2007) *Social Computing Upends Past Knowledge Management Archetypes*, Cambridge, MA, Forrester.

Brown, J. S. and Duguid, P. (1991) Organizational learning and communities of practice. *Organization Science*, 2, 1.

Brown, J. S. and Duguid, P. (2000) *The Social Life of Information*, Boston, MA, Harvard Business School Press.

Brynjolfsson, E. and Hitt, L. M. (1996) Paradox lost? Firm-level evidence on the returns to information systems. *Management Science*, 42, 4, 541–558.

Bulkley, N. and van Alstyne, M. W. (2006) An Empirical Analysis of Strategies and Efficiencies in Social Networks. Available from: http://ssrn.com/abstract=887406 [Accessed 16 October 2007].

Charron, C., Favier, J. and Li, C. (2006) *Forrester Big Idea: Social Computing*. Cambridge, MA, Forrester.

Constant, D., Sproull, L. and Kiesler, S. (1996) The kindness of strangers: The usefulness of electronic weak ties for technical advice. *Organization Science*, 7, 2, 119–135.

Davies, J., Duke, A., Kings, N., Mladenic, D., Bontcheva, K., Grcar, M., Benjamins, R., Contreras, J. and Blaquez, M. (2005) Next generation knowledge access. *Journal of Knowledge Management*, 9, 5, 64–84.

Duke, A., Davies, J. and Glover, T. (2007) Squirrel: An advanced semantic search and browse facility. In Franconi, E., Kifer, M. and May, W. (Eds.) *European Semantic Web Conference (ESWC) 2007*. LNCS 4519 ed. Innsbruck, Springer-Verlag.

Golder, S. and Huberman, B. A. (2006) Usage patterns of collaborative tagging systems. *Journal of Information Science*, 32, 2, 198–208. Available from: http://www.hpl.hp.com/research/idl/papers/tags/index.html [Accessed 8 October 2007].

Granovetter, M. S. (1973) The strength of weak ties. *American Journal of Sociology*, 78, 6, 1360–1380.

Granovetter, M. S. (1983) The strength of weak ties: A network theory revisited. *Sociological Theory*, 1, 201–233. Available from: http://www-personal.si.umich.edu/~rfrost/courses/SI110/readings/In_Out_and_Beyond/Granovetter.pdf [Accessed 2 October 2007].

Heath, T., Motta, E. and Petre, M. (2007) Computing Word-Of-Mouth Relationships in Social Networks from Semantic Web and Web2.0 Data Sources. *Bridging the Gap between Semantic Web and Web 2.0 Workshop*. 4th European Semantic Web Conference, Innsbruck, Austria. Available from: http://www.kde.cs.uni-kassel.de/ws/eswc2007/proc/ComputingWorld-of-Mouth.pdf [Accessed 8 November 2007].

Hinchcliffe, D. (2006) Creating Web 2.0 applications: Seven ways to fully embrace the network. *Social Computing Magazine*. Available from: http://web2.socialcomputingmagazine.com/creating_web_20_applications_seven_ways_to_fully_embrace_the.htm [Accessed 8 November 2007].

Hyams, J. and Sellen, A. (2003a) Gathering and sharing web-based information: Implications for 'ePersons' concepts. HP Laboratories, Bristol. Available from: http://www.hpl.hp.com/techreports/2003/HPL-2003-19.html [Accessed 8 October 2007].

Hyams, J. and Sellen, A. (2003b) How Knowledge Workers Gather Information from the Web: Implications for Peer-to-Peer File Sharing Tools. *HCI 2003: Designing for Society*. Bath, UK. Available from: http://www.hpl.hp.com/techreports/2003/HPL-2003-95.html [Accessed 8 October 2007].

Ichniowski, C., Shaw, K. and Prennushi, G. (1997) The effects of human resource management practices on productivity: A study of steel finishing lines. *American Economic Review*, 87, 3, 291–313.

Kings, N. J., Alsmeyer, D. and Owston, F. (2003) Libraries as Shared Spaces. *Universal Access in HCI: Inclusive Design in the Information Society. Proceedings of HCI International, 2003, Volume 4.* Lawrence Erlbaum Associates.

Kings, N. J., Gale, C. and Davies, J. (2007) Knowledge Sharing on the Semantic Web. In Franconi, E., Kifer, M. and May, W. (Eds.) *European Semantic Web Conference (ESWC) 2007*. LNCS 4519 ed. Innsbruck, Springer-Verlag.

Marshall, C. C. and Brush, A. J. (2004) Exploring the Relationship between Public and Private Annotations. *Proceedings of the ACM/IEEE Joint Conference on Digital Libraries (JCDL04)*, 349–357. Available from: http://www.csdl.tamu.edu/~marshall/pubs.html [Accessed 8 October 2007].

Marwick, A. D. (2001) Knowledge management technology. *IBM Systems Journal*, 40, 4, 814–830. Available from: http://www.research.ibm.com/journal/sj/404/marwick.html [Accessed 2 October 2007].

McKinsey (2007) How businesses are using Web 2.0: A McKinsey Global Survey. Available from: http://www.mckinseyquarterly.com/How_businesses_are_using_Web_20_A_McKinsey_Global_Survey_1913_abstract [Accessed 23rd October 2007].

Mika, P. (2005) Ontologies are us: A unified model of social networks and semantics. In Gil, Y., Motta, E., Benjamins, V. R. and Musen, M. A. (Eds.) *The Semantic Web – ISWC 2005*. Galway, Ireland, Springer.

Nairn, G. (2006) Social networking becomes work. *FT.com*. Available from: http://search.ft.com/ftArticle?ct=0&id=060411009110 [Accessed 8 October 2007].

Raybourn, E. M., Kings, N. J. and Davies, J. (2003) Adding cultural signposts in adaptive community-based virtual environments. *Interacting with Computers*, 15, 1, 91–107. Available from: http://dx.doi.org/10.1016/S0953-5438(02)00056-5 [Accessed 22nd October 2007].

Rogers, E. M. (1995) *Diffusion of Innovations*, New York, NY, Simon & Schuster Inc.

Shneiderman, B. (2003) *Leonardo's Laptop: Human Needs and the New Computing Technologies*, London, England, MIT Press.

Wenger, E. (1999) *Communities of Practice: Learning, Meaning and Identity*, Cambridge, MA, Cambridge University Press.

3

Device Futures

Jonathan Mitchener

3.1 Introduction

Understanding the future of devices is fundamental to understanding the future of the
ICT industry. The device is the way that a customer connects to networks, and shapes
the way that customers experience services and applications [in this chapter, the term
device is used to cover a wide range of equipment – from mobile phones, personal
digital assistants (PDAs), laptops to smaller ambient devices that may be embedded
in the environment]. The device market also happens to be the focus of most of the
innovation in our industry. Competition in the markets is fierce, with only a minority
of launched products becoming widely used.

This chapter has been organized in a series of short sections:

- *New Modes of Interaction* examines possible developments in the ways that people
  interact with devices.
- *Device Component Trends* surveys the technology trends associated with the major
  components used by devices.
- *Broader Device Trends* lays out some of the challenges associated with device
  futures.
- *Information Anywhere, Anytime* looks at the consequences of device evolution.
- *Ambient Information* speculates on the way that ambient information will be used.
- *Intelligent, Context-aware Interaction* reviews the impact of increased intelligence.

3.2 New Modes of Interaction

Devices of the future will allow users to interact with them in new ways. Continued
improvements in processor speeds will mean that ever more processor cycles are available to interpret and to provide more natural and sophisticated user interactions.

ICT Futures: Delivering Pervasive, Real-time and Secure Services
Edited by Paul Warren, John Davies and David Brown
© 2008 John Wiley & Sons, Ltd

Speech recognition and synthesis technology has advanced significantly in recent years, building on the previous decade's progress (Cole and Zu, 1996). In closed environments, such as vehicles, and in applications where a limited vocabulary is needed, talking to a computerized device to interact with it can be natural and effective. Some cars, for example, now have quite complex dialogue structures enabled by speech recognition such that the driver may have a significant 'conversation' hands-free about some aspect of the car's facilities (e.g. telephone, navigation or audio system), rather than single keyword commands. Speech synthesis for receiving information from a device can also be appropriate when the output can be broken up into small quantities and in environments where there are no privacy or interference issues to be concerned about. Speech output need not be robotic sounding as the historic stereotype suggests.

Haptic interaction (by which we mean touch-based feedback between user and a device) is still relatively rare between machines and their human users. The most obvious examples are in the vibration mode of many current mobile phones, and its use in action game playing such as flight simulations for instance. More general game consoles have begun to employ wireless implements that can be swung, pushed or otherwise moved to mimic the relevant action for the game in question, such as swinging a tennis racket or brandishing a sword! Haptic output may initially be seen in combination with otherwise 'flat' touch screen buttons in order to give the user some subtle feedback that the button press has been registered and acted upon.

The touch screen is finally becoming a mainstream input/output method on phones and PDAs, and some of these are moving beyond the simple stylus pointer (Hinckley and Sinclair, 1999). Once the touch screen is accurate enough to allow the use of human fingers, a whole world of more complex gestures is opened up, expanding the set of metaphors that can be used (pinching movements to change the size of objects on the screen, for example). The recent introduction of the Apple iPhone shows how an effective touch interface can be successful in the market. This is likely to develop into a de facto set of such gesture standards analogous to the mouse point and click metaphor set that has become so commonly understood today within the personal computer windowing user interface (UI) environment. Gestures with fingers on this sort of screen might include tap, tap and hold, tap and drag, double tap and stroke. All these gestures can be done with a single finger. However, unlike devices intended for a pointer or stylus, those intended for human fingers can also easily accommodate gestures involving multiple fingers, pinching being one example among many.

Tilting and otherwise moving the whole device will also feature on some gadgets, perhaps especially those aimed at users with specific dexterity problems. An example is BT Balance (Figure 3.1). The speed and angle of movement, tracked by accelerometers, can be used to signal different intuitive actions such as tipping back and forth quickly to advance to the next page of information, for example. Devices of this type thus begin to mimic the natural actions one might take when reading a traditional paper book. Users tend to be positive about interaction methods that are more natural than the traditional mouse and keyboard (Kolo and Friedewald, 1999). Mobile phones using accelerometers are used in Japan, and, when combined with location information, allow mobile phones to point to buildings and allow information to be provided on very specific locations.

Figure 3.1   BT Balance (Reproduced by permission of BT)

Further into the future, we might expect facial recognition to form part of the way one can interact with a device (Shell *et al.*, 2003). The face not only provides key identifying features but can also suggest the mood and intent of the user as well as indicate whether they are paying attention to the device. The future will see closer interaction between human and machine (Figure 3.2).

3.3 Device Component Trends

Looking at the trends of the component technologies from which devices are made gives a strong indication of where devices themselves may be headed. This is true whether the components are processor chips, storage, network and graphics adaptors, or power sources such as batteries.

3.3.1 Processors

Moore's Law (Moore, 1965) is continuing unabated at present. Various techniques in processor design, including smaller die sizes, the addition of huge cache memories and multiple cores on a chip, have meant that performance marches on, with lower power consumption and smaller physical sizes. Specialized processors for graphics and networking can also offload some work from the main processor freeing up cycles. As a result of many of these trends, more powerful processors are finding their way into much smaller handheld devices, which in turn means that more sophisticated UIs, and more powerful operating system software, can be supported on such devices.

3.3.2 Storage

Relatively cheap, physically miniaturized storage capacities have grown massively in recent years. The general trend of capacity and size over time applies to practically all technology types, whether optical, magnetic or electronic. Advanced techniques in

Figure 3.2   Future human–machine interaction (Reproduced by permission of BT)

magnetic disks such as perpendicular recording have pushed capacities to many hun dreds of gigabytes per square inch. Many new technologies are being explored in research labs. Approaches such as anti-ferro magnetically coupled layers, micro electro-mechanical system probe devices, then holographic and eventually atomic level storage will be developed. The nanotech industry is already touting carbon nanotube memory as a faster, denser and eventually cheaper future replacement for solid-state memory chips (Rueckes *et al.*, 2000). More storage opens up new applications. Today's portable music players are giving way to devices that easily handle video and multimedia.

3.3.3 Network Adaptors

Traditionally, most devices support a single network type. Devices are starting to support multiple networks – for example, mobile devices are now appearing that support cellular, Bluetooth and Wi-Fi network connections. In the future, devices will support multiple network types and switch seamlessly between them according to the current context and any specified policy regarding preferred networks and other preferences (for example, involving cost and quality). Maintaining more than one connection at a time and switching intelligently according to the task in hand will become important characteristics. Software techniques (Mitola, 2000) for accommodating the myriad of wireless technologies will be increasingly important where both spectrum and modulation methods vary according to evolving standards. Wired connections are always likely to lead wireless connections in terms of raw maximum speeds attainable, and reliability advantages may be important in applications where mobility and portability are not.

3.3.4 Power Sources

Power source evolution has moved slower than other forms of components. The battery still dominates, although the exact chemistry involved has varied over time. There are three methods by which the power lifetime of devices is likely to improve in performance in the future. The first is the continued evolution of battery technology itself. The second is the employment of alternative sources such as fuel cells, solar or mechanical methods. Among these alternative sources is *power scavenging*, made possible by the use of intelligence. This radical alternative involves devices that are able to generate power from a variety of sources in their immediate environment including, for example, sunlight, changes in temperature, vibration, motion, sound and pressure. The environmental and consumer benefits are clear. Power-scavenging technologies are most actively being used at present in radio frequency identification (RFID) tags and in remote sensing technologies, as these small-scale devices have limited and often temporary power requirements.

The third method for improving the power lifetime of devices is more intelligent power management by the device itself, which can be achieved in various ways, including the automatic seamless turning off and on ('regulation') of components within the device as needed. In addition, techniques such as inductive charging (mainly exploited by electric toothbrushes at present) may make recharging batteries in devices more convenient.

3.3.5 Displays

While innovation in power has been relatively slow, display technology continues to move forward at a pace. The display typically represents one of the greatest costs of devices. The cost of liquid crystal displays (LCDs) is going down while new lower power and more flexible (sometimes literally, with electronic paper, for example) alternatives are appearing. Larger screens are becoming widespread. As mentioned before, screens will increasingly be used for input as well as traditional output.

3.4 Broader Device Trends

3.4.1 Co-operating Devices

At present, the gadgets we buy as consumer electronic goods are generally limited in terms of their ability to interwork. They form 'islands' of capability (e.g. the TV and its set-top boxes, your computers, your mobile phone and PDA, your camera and camcorder, etc.). With (sometimes considerable) user effort, it may be possible for one device to use the internet connection of another, or to synchronize the information stored on one with that on another. Using a bigger or better screen on one device for the display of data from another is much harder.

This is an area where there is considerable opportunity to improve usability, to enable new innovative services, and to provide a route to convergence rather than a 'one-size fits all' approach of bundling as much functionality as possible into one physical package. Users understand the key capabilities of individual devices. They also understand the value of relatively complicated services that could stem from the interaction of a group of co-operating devices. Users do not – and should not need to – understand the complex details of how individual devices communicate. This is convergence by sharing; the value of the sum of the parts is greater than the value of the individual components.

One challenge is that suppliers focus on differentiation, and this reduces the incentives for manufacturers to work together to achieve tighter device co-operation. But as markets become more mature, companies often choose to collaborate more. Already internet enthusiasts are promoting open interfaces and standards (so many mobile phones can run software applications that enable other devices to be controlled from the mobile phone). In any case, the days when the average user had just one device from one manufacturer are over: one recent survey found that 79% of those surveys carried typically carried at least two devices with almost one in five saying they carried four or more.[1] So there is potential for product or service differentiation by enabling co-operation and interworking between devices, particularly if one takes into account the user experience dividend which users really value. And there are examples of groups forming that are planning to make this kind of co-operation real (like the Open Handset Alliance[2]).

Figure 3.3 shows the trade offs between convenience and complexity that manufacturers have to make when they think about convergence. Bundling more features into a single device only pushes the cost beyond more potential purchasers and unavoidably adds to the complexity of the device. Allowing simpler optimized devices to use the capabilities of others in a seamless co-operative way opens up the market more and masks the complexity of the whole 'system'.

3.4.2 Masking Device Complexity

Underlying component trends make it clear that devices will become more complex, and probably have increased functionality over time. Suppliers' drive for differentiation also means new features and designs are continuously entering the market. This often makes devices harder to use. There are many people who find existing devices difficult to understand and use, so there is a real incentive for suppliers to mask this increasing complexity from the user, allowing them to focus on the task at hand, rather

---

[1] http://management.silicon.com/itpro/0,39024675,39129263,00.htm.
[2] http://www.openhandsetalliance.com/.

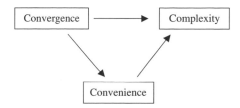

Figure 3.3   The complexity triangle

than being distracted by the technology used to carry out that task. Companies that are good at doing this (for example, Apple with the iPod and iPhone) can charge significant premiums for the devices that they sell. It seems likely that more companies will follow the strategy of using good design and simplicity strategies to capture larger market shares (an example of this would be Philips, whose marketing campaigns currently focus on simplicity).

Much has been written regarding the problem of making the user experience easier and better, whatever the system, service or device (Fu *et al.*, 2004; Norman, 2007). Devices are the way that users experience services. A great deal of effort often goes into UI design, both for traditional and newer modes of interaction (see Section 3.2). While this is important, there is only so much simplification that can be achieved for a particular task or function within the UI. Hence, more holistic approaches to simplification are going to be necessary as services become richer and devices become ever more capable. The co-operative approach between devices offers one such way to mask the complexity of operation from the user's perspective.

Another trend in devices is that formerly disparate device types (TV, computer, phone) can all increasingly be seen as simply a computer platform that has been optimized for a particular purpose and has been encased in suitable aesthetic materials to appeal to a specific market. This notion of generic computer platforms being specialized by the addition of mainly software components to turn them into a specific purpose product is very powerful but carries with it the danger of making the complexity of every device that of the personal computer. It is one thing to have to maintain the software for the operating system of a personal computer; it is quite another to have to do the same for a TV, something which is now a reality for some users.

3.4.3 Always-connected Devices

With the widespread take-up of services like broadband, people are becoming used to the idea of always being connected, at least in the home or in the office. On the move, companies like RIM offering products like the Blackberry have achieved considerable success with an 'always-connected' product. But current products are often expensive, and 'always connected' has yet to spread widely to other classes of device. Issues of coverage, cost or a painful manual connection process still need to be overcome. The always-connected devices of the future will manage this much better on behalf of the user and will have to be accompanied by sensible business models that free the user from concerns about whether being connected in a particular location or via a particular network is affordable.

At that point, just as with fixed internet applications, take-up will explode. In the fixed broadband world, the use of online shopping sites, social networking applications and simple video chatting with family and friends from home has mushroomed since payment schemes switched from pay-per-minute to a flat rate model. Consumers value fixed cost packages and the more relaxed experience that come from not having to worry about variable costs. When this is also true for mobile connectivity, mobile data usage will grow and a corresponding range of new applications will flourish.

Devices will have a range of networks to choose from in future. The question will not be whether a device can connect but via which network it should connect. This is not a simple choice of the fastest connection taking priority, but balancing the needs of the user and applications being used against the price sensitivity of the customer. Price premiums are likely to remain in relation to connection speeds, and so a slow cheap connection might well be the optimum choice by the device if the application in progress is, for example, a low volume information synchronization activity. However, should the user suddenly decide to begin a more data intensive application, then the switch to a more expensive but higher bandwidth network connection should be as seamless and transparent as possible.

3.5 Information Anywhere, Anytime

The always-connected device of the future will bring the ever-increasing power and utility of the Internet to its users wherever they are at anytime. People will be able to get information, do business and be entertained anywhere at anytime. This section discusses this point from, first, the user perspective and then from the device angle, highlighting in each case how device intelligence will help create a better experience.

3.5.1 The User Dimension

While the potential to stay informed and to access services ubiquitously may be good, users also need to manage the threat of continuous interruptions and to prevent information overload. There are already concerns in society about work–life balance as well as the time that people now spend sitting and surfing as opposed to more physical and social activities in their free time.

Devices will need to learn when the user is likely to want to be contactable and by whom, and in which places. This is a complex task requiring maintenance of a detailed profile of the user including information about what they are doing, where they are doing it and with whom. Information may come not only from the device, but also from sensors and other devices in the environment. This level of knowledge raises big security and privacy issues. Availability of information about individuals can be used for the good (for example, allowing parents to track the location of young children) or for evil (allowing stalkers to track targets).

The suppliers that will win will be those who succeed in developing a trusted relationship with customers. Users will want to remain in control, but one can envisage
deals being done – for example, the ability for advertisers to target mobile devices in
return for some benefit to the end user – for example, a cost-reduced service or
device.

3.5.2 The Networked Device Perspective

The always-connected device has scope to benefit from intelligence in the way it connects to different networks as mentioned in Section 3.3.4 but also in terms of what it
does at any particular moment. As is the case with fixed broadband, mobile devices
that are always connected to the Internet can perform a number of useful types of task
in the background when the user need not be aware. The challenge for personally
carried devices is to do this within additional constraints such as power consumption
control, for example. Examples of such background activities could be system/firm-
ware updates, information synchronization with other devices or information feeds on
the Net, backing up user data to a network store, or prefetching content that the user
is likely to request next. The latter could be done as a result of a previous instruction
by the user or anticipated by the device based on user preferences, previous actions
or circumstances of the time, place, etc.

3.6 Ambient Information

So far, in this chapter, we have considered scenarios where users actively engage with
devices in order to access information or to be entertained, for example. There is an
alternative paradigm that is more subtle but can also fit very snugly around this more
traditional model. The use of ambient devices allows information to be presented in
a very lightweight fashion as a background activity but in a way that allows the user
to be informed without feeling overloaded or bound to seek out the data themselves.
Essentially, an ambient device is less intrusive and 'glance-able', utilizing so-called
pre-attentive processing (Treisman *et al.*, 1992).

A number of ambient devices have appeared in recent years including orbs, dashboards and beacons. Such devices may take the form of ornaments, plants or other
objects (e.g. the Nabaztag rabbit) (Peters, 2006) within an environment. They typically
convey information through the use of colours, physical positioning or signals, and
perhaps subtle sounds. Even digital photo frames might be classed as a very crude
form of device that can offer ambient information, especially if it is not simply photo-
graphs that are displayed. Ambient devices are particularly suited to presenting information that represents status (such as the number of new emails in your inbox, or the
number of friends online at a specific time), or that is denoting a trend of some sort
(such as weather parameters, stock prices or availability).

Many internet users have a number of email accounts, blogs, wiki sites, social networking sites and news feeds to look at periodically, sometimes daily or more in addition to their offline world of phone voicemail, answer phones, and TV and radio news
content. Signaling some of these using ambient approaches may offer some respite!

Indeed in the future, probably the most compelling use of ambient information will be to supplement rather than to replace other forms, with ambient devices being an integral subset of the typical devices that people have around them. In a similar vein to the device co-operation described in Section 3.3.2, the maximum value will be derived from the perhaps complex but seamless interaction between ambient and other devices which results in additional simplicity for the user.

3.7 Intelligent, Context-aware Interaction

This chapter began by looking at modes of interaction between people and machines, specifically computer-based gadgets and devices. The device trends section has indicated that the march towards faster, more powerful devices, capable of storing more and connecting to more networks than ever before is looking relentless. And the latter sections have covered future ways of accessing information and have highlighted the benefits of applying intelligence to this issue. This final section, therefore, draws on all of these points to imagine a world where future devices are able to exploit some notion of intelligence to decide for themselves the most appropriate way to interact with their human 'operators' or 'counterparts', according to the context of the situation.

A simple example of context-sensitive device behaviour would be a mobile phone that can switch off its own audible ringer and adopt the silent vibrate mode automatically when its owner has entered a meeting, or public gathering such as a cinema or theatre. Similarly, it would be useful if a PDA is able to switch to voice prompting about a calendar reminder if the owner is currently driving his/her car.

With more processing power, software intelligence will be applied to human–machine interaction. It seems likely that we will see intelligent systems deployed far more widely. One example is the development of the robot marketplace. Expensive robots are widely used on automated car production lines welding metal panels together. In Japan, and increasingly in other markets, robots are being developed for the domestic market where much more human interaction is required. With demographics pointing to a future for some communities where there are insufficient younger people to support the larger number of older generations, in applications such as healthcare and monitoring, robots may have a key role in the future. Other quite obvious applications include security, cleaning, hosting and directing as well as social equivalents to pets. In all of these situations, it is almost imperative that the machine exhibits some intelligence with regard to suitable interactions taking into account its context.

So what aspects of context are there to consider? Well nothing should be ruled out; indeed, as humans, we are very good at taking into account all sorts of contextual information as we go about our lives. For a machine to do likewise is very ambitious, given that some element of each context has to be modeled in some way first. But by bringing together a large number of simple facts about the environment in which a robotic device finds itself, a significant contextual picture can be derived. The contextual picture might include where, when and what scenario is taking place, how many other people are present, whether the ambient noise level is high or low, information about the environment such as weather, what plans are filed (e.g. in a diary) and a detailed profile of the person with whom the interaction is to take place. From an analysis of this, the appropriate interaction, in terms of mode, style and content, can be chosen.

3.8 Conclusion

This chapter has shown that the future of devices and gadgets is an exciting one that will open up more opportunities for users. New modes of interaction will offer more natural human methods for using devices as a result of more processing cycles being available. This allows sophisticated interactions to take place as a result of the machine correctly analyzing a myriad of complex signals given by the user. These signals may be in the form of gesture, speech or action, for example.

In the future, network connections will be more numerous, automatic, seamless and simultaneous. Increased co-operation between devices will not only increase utility but will mask the consequent increased complexity from users, an issue of critical importance for user acceptability. There are also sound business reasons to make this possible through standards and interworking. The last overall trend identified is the idea of many devices simply regarded as generic computing platforms that can be specialized according to the software build applied. This is a notion that brings flexibility and potential, but could also bring some of the general computer complexity to other devices which should actually be far more obvious and easy to use.

The always-connected device brings advantages for users, but also raises privacy and security issues that will need to be dealt with. Increased intelligence brings easier-to-use devices, but also is likely to increase the use of ambient devices embedded in the environment, and to lead to broader use of robotics in the consumer marketplace.

References

Cole, R. and Zue, V. (1996) Spoken Language Input, Chapter 1, in Cole, R. *et al. Survey of State of the Art in Human Language Technology*, Cambridge, Cambridge University Press.

Fu, R., Su, H., Fletcher, J., Li, W. *et al.* (2004) A framework for device capability on demand and virtual device user experience, *IBM Journal of Research and Development*, Sep–Nov.

Hinckley, K. and Sinclair, M. (1999) Touch-Sensing Input Devices. *Info on Proceedings of the ACM CHI 99 Human Factors in Computing Systems Conference*, pp. 223–230, Pittsburgh, PA, ACM.

Kolo, C. and Friedewald, M. (1999) What Users Expect from Future Terminal Devices: Empirical Results from an Expert Survey. *Proceedings of UI4All.*

Mitola, J. (2000) *Software Radio Architecture*, John Wiley & Sons, Ltd.

Moore, G. (1965) Cramming more components onto integrated circuits, *Electronics*, 38, 8, 19 April.

Norman, D. (2007) *The Design of Future Things*, New York: Basic Books.

Peters, L. (2006) Nabaztag Wireless Communicator, *Personal Computer World*, 02 May 2006.

Rueckes, T., Kyoungha, K., Joselevich, E., Tseng, G., Cheung, C. and Lieber, C. (2000) Carbon Nanotube-Based Non-volatile Random Access Memory for Molecular Computing, *Science*, 289 (5476), 94–97, 7 July.

Shell, J., Selker, T. and Vertegaal, R. (2003) Interacting with groups of computers, *Communications of the ACM*, 46, 3, March.

Treisman, A., Viera, A. and Hayes, A (1992) Automatic and pre-attentive processing, *American Journal of Psychology*, 105, 341–362.

4

Online Trust and Customer Power: The Emergence of Customer Advocacy

Glen L. Urban

4.1 Introduction

Trust has been recognized as a key component in determining the successful impact of the Web ever since its inception. As the popularity and usage of the Web continues to grow, security and privacy of online transactions brings to light the necessity for trust. At the same time, the Internet has drastically increased customer power. Buyers gain new information, experience new intermediary firms, and have simplified transactions. Travel is a good example where it is easy to get information on rates and schedules with sites like Travelocity, Expedia, and Orbitz. This gives customers the power to find and buy the lowest cost ticket (or shortest travel time if that is their need) and to get the eticket directly without the involvement of a travel agent, therefore avoiding the travel agent fee. Similarly, over 70% of people in the developed world get car information over the Internet and visit dealers with the power of knowing invoice prices, discount promotions, and bargaining targets. The same story can be told for stock buying and, of course, consumer durables where sites like Amazon and ePinions represent new intermediaries, and existing stores like Wall Mart and Sears have sites that offer full information to consumers.

Today, trust and customer power have partnered to revolutionize marketing. Marketers and IT managers are challenged with the task of changing the online climate in order to gain and retain online consumers. This has generated tremendous interest in learning about online trust and in developing new site designs to respond to the increased power of customers.

ICT Futures: Delivering Pervasive, Real-time and Secure Services
Edited by Paul Warren, John Davies and David Brown
© 2008 John Wiley & Sons, Ltd

I begin this chapter by examining the state of the art of online trust and by reviewing a selected set of findings from the past research that describe what we know about online trust. Then, after identifying some innovative sites now on the Web that are especially effective in building trust, I outline a series of projects conducted at Massachussets Institute of Technology that develop a customer advocacy methodology that builds trust and responds to the new magnitude of customer power. In contrast to pushing harder with existing marketing tools and improved targeting, this emerging consumer advocacy methodology partners with customers and helps them make the best decision for their needs by supplying full and honest information and advice on all products. I describe research project work at General Motors, BT, and Suruga Bank. This chapter closes with a look at the future and a prediction that a growing number of firms will use IT to support a trust-based customer advocacy strategy for their firms.

4.2 What is Trust?

Trust in the traditional offline market has been examined from multiple disciplines, and naturally, different definitions of trust emerge within each discipline. These studies have contributed significantly to the study and application of online trust. I like best the definition by Rousseau *et al.* (1998) where he defines trust as 'a psychological state comprising the intention to accept vulnerability based on positive expectations of the intentions or behaviors of another'. Trust is an intermediary attitude evoked when a customer believes that another (person or company) will act in his or her interest when the customer faces the vulnerability of an adverse outcome.

Using this definition of offline trust as a starting point, researchers have widened the definition of trust and applied it to online trust. Online trust includes consumer perceptions of how the site would deliver on expectations, how believable the site's information is, and how much confidence the site commands. In essence, trust is developed when consumers form positive impressions of an electronic merchant's site and are willing to accept the vulnerability of making a bad purchase decision.

In practice, advertisers also recognize the critical issue of vulnerability and some define trust as 'not being afraid when you are vulnerable' (Saint Paul Insurance Co., 2000). The definition of trust has evolved over the past decade and, although semantics may differ, has now reached general consistency among researchers. Trust can be distilled down to three component dimensions: integrity/confidence, ability/competence, and benevolence. If you have all three, trust will be high.

4.2.1 Trust is More Than Privacy and Security

When the Internet was in its infancy, privacy and security were critical elements that online businesses addressed to earn consumer online trust, and they were often cited as antecedents to trust. However, with the maturing of the Internet, consumers have come to expect more from online businesses, and their requirement for trust has also increased (Shankar, Urban and Sultan, 2002). Privacy and security have become the new baseline from which one evaluates an online merchant's trustworthiness. While progress has been made in improving security, we continue to see viruses, identity theft, and phishing. Vigilance is required and must be evident in sellers' site efforts –

privacy and security are assumed, and so this expectation must be fulfilled or there will be a 'trust buster'.

Brand is an important conveyor of trust that extends beyond privacy and security. Branded products with strong brand equity enjoy an immediate trust gain when in an online environment. Past experience through order fulfillment is an important determinant of the trust consumers have in the site. Good execution of the purchase is expected, and if this expectation is not met, trust will decrease.

Another dimension that influences trust is online peer and editorial recommendations. Essentially, these recommendations empower consumers to make decisions based on availability of information. Online peer recommendation is an invaluable resource for consumers, and many do use peer recommendations to make purchasing decisions. The provision of reviews builds confidence in the mind of the consumers and subsequently, the consumer proceeds with a purchase. In this environment, transparency and trust are strategic dimensions that are reflected in visual communication, advice, and information content.

4.2.2 Site Design Affects Trust

While the privacy and security statements of electronic merchants are appreciated by consumers, it is ultimately the Web site's design that influences consumer trust and consequently impacts online purchase intentions. Consumers make intuitive, emotional decisions based on their perception of an online merchant's Web site. The look and feel of a Web site serves as a basis for consumers to form a first impression of the merchant and thus forms an opinion of his trustworthiness, and can affect their probability of purchasing.

In a recent marketing study of site design features and trust, Bart *et al.* (2005) found a number of site variables that affect consumers' trust. They reconfirmed that although privacy and security are important, they are not as important as user-friendly navigation and presentation. After brand and navigation, the next most important site design aspect is the perception of a seller's 'assistive intent' through interactive tools. This plays an important role in the formation of online trust. In essence, the fact that an online seller has tools that help consumers process and manage the plethora of product information on its Web site will lead to the belief that the seller has the consumer's interest in mind, and therefore the consumer is more willing to trust the online merchant, especially in the case where the online merchant is new to the consumer. When consumers feel empowered, trust is formed.

It may be obvious, but errors in the site (missing links or pages and inconsistencies) are trust busters and must be avoided by careful quality assurance. The prevalence of site errors is decreasing as standards and testing improve along with better privacy and improved security. But with increasing complexity and frequent changes, errors can creep into even the best Web sites if quality assurance procedures are not at the highest possible levels.

A growing number of studies show the positive correlation between the perceived visual communication quality of a Web site's user interface and overall user satisfaction and trust. In a study with over 2600 participants, the Persuasive Technology Group (Fogg *et al.*, 2003) at Stanford University evaluated the credibility of different Web sites and gathered the comments written by participants about each site's credibility. They found that the design of the site is mentioned most frequently, being present in 46.1% of the comments. Site design from both an IT and a marketing perspective is important in generating trust.

4.2.3 Trust is a Process

While the specific site design for an online visit is important, it should be realized that it is rare that trust is built in one session; usually trust is the result of a process of gaining experience and increasing trust as expectations are met. Emerging research has started to evaluate trust from a process-centric perspective – initial trust and continuing experience are important.

In particular, initial trust is built when a consumer first visits a new online merchant's site. The primary factor that builds trust among new consumers to a Web site is the online merchant's ability to empower the consumer to understand and manage product information and to complete a given task. In the beginning of the relationship, when no priors have been established, consumers form trusting beliefs based on cues utilized by online merchants.

Consumers' buying behavior is influenced by the trust initially developed as they interact with a Web site, but repeat visits and purchases are just as important. Trust is developed when consumers have a positive experience with an online merchant through such things as order fulfillment, service, product satisfaction, or reputation of the online merchant. Trusting beliefs are confirmed when online merchants fulfill the order and provide exemplary customer service; trust is also reinforced when the customer experiences satisfaction with the product over time. Trust is difficult to gain, but is easy to lose. Continuing trust is an important component in the buyer–seller relationship that needs to be cultivated and maintained over time by implementing a rigorous quality control program.

With this basic understanding of online trust, next we can examine the impact of the growth in customer power and begin to define strategies to deal with this new online environment.

4.3 What to Do About Growing Customer Power?

While trust retains its important status as a necessary criterion for effective marketing on the Internet, customer power growth has amplified its effect. We do not have to look far to see evidence of this power. Earlier, we cited that over 70% of consumers go to the Internet for auto information and travel information. But this number is generalizable to financial services, real estate, and even health services. In almost every industry, customers are gaining more information on purchase alternatives that allows them to make better decisions and insulates them from less than transparent marketing efforts. Eighty percent of customers say they will not buy from sellers they do not trust. Sales on the Internet in the USA will total over $200 billion this year, and it is clear that trust is the currency of customer power.

Other technologies also enable increased customer power. Over 145 million numbers in the USA are on the 'no call' list. Eighty percent of people switch channels or mute ads and TiVo and DVRs are allowing customers to fast-forward over ads. Customer complaint sites and customer ratings have become more prevalent. The impact is evident on eBay where ratings of used car sellers have forced successful sellers to be honest and complete in the information they provide – if they are rated badly, few potential customers will buy from them. eBay has taught previously disreputable used car sellers to be honest if they want to sell successfully on eBay. Over $5 billion dollars in used cars are sold on eBay each year and only by honest and highly rated sellers.

What is a company to do about this requirement for trust and increase in customer power? One answer is to fight harder – to do more one-sided advertising and heavy-handed promotion and to increase efficiency with key word advertising and individual ad targeting. Increased advertising in mass media, even while audiences and effectiveness decline, demonstrates many firms are trying this 'push harder' strategy. This may work in the short run, but I believe the fundamental force of customer power will overwhelm such 1950s style marketing.

There is a better answer – partner with your customers and help them make the best decision based on open and honest information. I call this 'Customer Advocacy'. Some firms are being successful with this type of strategy. See Figure 4.1 for an example of progressive insurance where transparent competitive auto insurance quotes are provided on the Internet, and Progressive even supplies the URL and phone number for its competitors. Another example is Experion Systems which provides a 'plan prescriber' to educate and advise senior (over 65 years of age) consumers on the best prescription insurance for them (see Figure 4.2). Similar advisors are available on sites like MyProductAdvisor.com, which gives transparent advice on autos, cameras, and TVs.

A recent book by Urban (2005) called 'Don't Just Relate – Advocate' describes in detail the growth of customer power, and his customer advocacy pyramid is a guide to this new strategy implementation (see Figure 4.3). Firms build on a foundation of total quality and customer satisfaction to build trust consistent with the dream of Customer Relationship Manager to build a new transparent partnering relationship with customers. At the top of the pyramid, firms move to advocacy where they provide honest open advice *and* competitive offerings to the firm's own offerings. This is based on the proposition of 'if I advocate for you, you will advocate for me'. In other words, it is a realization that customers have the power, and the firm must aim to satisfy customers by providing them with the best products and honest information to help them make the best choices. Customers will then tell others about the firm and help the firm to design better products. The IT challenge is to implement the customer advocacy pyramid on the Internet and to coordinate it with the firm's traditional distribution channels.

4.4 New ICT Frontiers in Customer Advocacy

The notion of a customer advocacy strategy requires some innovative IT efforts. In this section, I describe three research projects conducted at MIT to develop the methodologies of trust building and customer advocacy. First, a 3 year project with GM on auto buying and advocacy has shown that consideration, preference, and sales of autos can be increased by supplying buyers with full competitive information (e.g. specifications, brochures, and reviews), advice, communities, and competitive test drives.

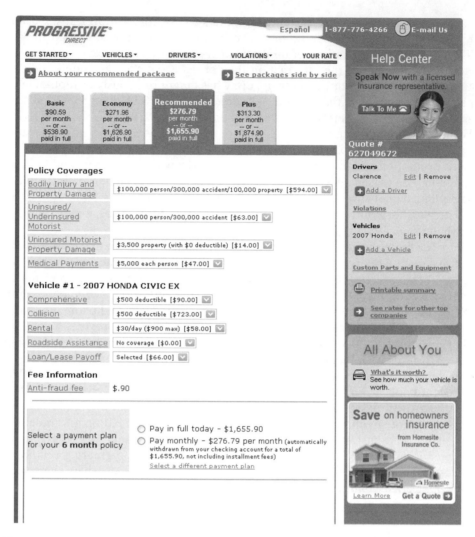

Figure 4.1   Progressive insurance (Reproduced by permission of The Progressive Group)

Second, a BT project extends the basic advocacy architecture by studying how one can morph the site to match a consumer's cognitive style. Finally, a Suruga Bank project generalizes the cognitive morphing to matching sites to culture and developing a global site strategy that adjusts an advocacy backbone to within and across country cognitive and cultural styles.

4.4.1 My Auto Advocate

MIT with support from GM has developed an advanced opt-in advocacy program called 'My Auto Advocate'. Figure 4.4 shows the first page of an interactive ad which

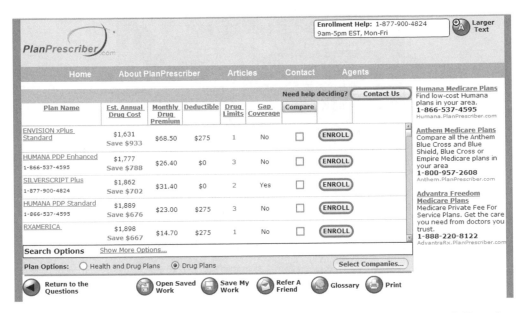

Figure 4.2 Experion Systems plan prescriber (Reproduced by permission of Experion Systems)

serves to convince people of the advantages of the system, and describes the unique program features. It includes, along with a customer community called a 'drivers forum', a virtual auto show with an advisor, competitive brochures, and test drives that fairly compare major GM and competitive models. Figure 4.5 shows the site built around a metaphor of an auto show. The virtual auto show is an interesting site design

Figure 4.3 Customer advocacy pyramid (Reproduced from Urban 2005 © Pearson Education)

Figure 4.4   My Auto Advocate opt-in interactive ad (Reproduced by permission of Glen Urban/MIT)

because it allows 360 degree panning and uses hot buttons to cue customer conversation bubbles as well as the major comparative information sources. The site also contains research on cars via the National Transportation Health and Safety Administration and the Insurance Institute that gives customers information they would not get very easily. Figure 4.6 shows the competitive test drive feature of the site.

One unique aspect of the site is the reward structure. The customer is paid through Amazon Coupons up to $20, to visit the advisor, test drive, participate in the community, or read eBrochures. In this time of media saturation and decreasing effectiveness, this payment for considering information may be the wave of the future for marketing and advocacy. People are paid for 'doing their homework', so they make a good decision.

A customer advocacy site can empower customers and cement the trust relationship with the sponsoring firm if the firm is truly the custodian of the customers' best inter-

Figure 4.5   My Auto Advocate home page (Reproduced by permission of Glen Urban/MIT)

est. Sites will honestly provide full information and provide advice on their products and competitors. They will have extensive educational materials in learning centers, and will host community dialog and feedback in dedicated social networks. The auto advocacy system described here was tested with over 5000 customers in Los Angeles to determine the impact of the components in an experimental and then in an opt-in study. Both these market research studies and an actual implementation at a dealer (Coulter Cadillac in Phoenix Arizona) have shown that such an advocacy can increase the number of people who will consider a GM vehicle, sales, market share, and profit. I believe advocacy sites similar to Figures 4.5 and 4.6 are a precursor of the future.

4.4.2 Customer Advocacy and Site Morphing to Match Cognitive Style

MIT also built a system of customer advocacy for broadband sales in the UK sponsored by BT (see Figure 4.7). Consumer research shows that consumers who visit this site significantly increase their consideration of BT and improve the probability of buying a BT broadband service. This market research indicates that transparency and honest

Figure 4.6 Auto show in motion (Reproduced by permission of Glen Urban/MIT)

advice to help consumers is the best approach, and it can have a positive effect on firms as well as customers even in markets that have been characterized by push marketing and less than full information. This study confirms the GM study conclusions that customer advocacy can increase sales and profits.

This BT study has extended the trust building aspect of customer advocacy by morphing the site to better fit individual cognitive style. Morphing means dynamically revising the site experience in real time based on customer clicks. The clicks reveal the customers' style.

In this methodology, an individual's decision style is modeled as analytic versus holistic, impulsive versus deliberative, or visual versus verbal. The site presents alternate levels of information and technical content. For example, if a person is analytic and impulsive, he/she will receive succinct data on products and will be able to make

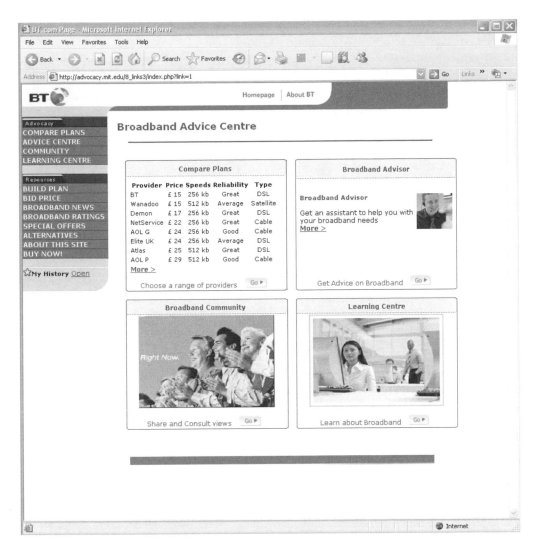

Figure 4.7   BT customer advocacy system (Reproduced by permission of Glen Urban/MIT)

a purchase in a minimum number of clicks. On the other hand, if he/she is holistic and deliberative, he will receive overall advice and detailed data so he can make an informed decision. Figure 4.8 shows how the BT site can change to match decision style, and supports a third dimension of visual versus verbal morphing. The morphing system uses Bayesian updates based on site clicks to infer the cognitive style of a user and optimally serves up the best morph for the consumer using machine learning theory. The proposition is that we trust sites that 'think like' we do. As in life with real people when we are 'on the same wave length', we believe and trust them more. This is just one example of how behavioral, marketing, and IT concepts can be integrated to improve the effectiveness of Internet trust strategies.

Morph 1   Morph 8

Figure 4.8 Side by side comparison of BT's customer advocacy and morphing system (Reproduced by permission of Glen Urban/MIT)

4.4.3 Cultural Morphing

Another research frontier is building flexible sites that capture within and across country differences. Hofstede (1980) has shown that there is as much variation within countries as across them, so the concept of morphing by cultural characteristics is attractive. If one has a worldwide site with a common backbone that then morphs within and across countries, individual differences could be fully captured. Currently, work is underway at MIT to extend the system for cultural dimensions such as egalitarian versus hierarchical, individual versus collective, and emotional versus neutral. For example, Figure 4.9 shows a backbone site for Suruga Bank in Japan. While much of Japan is traditional (hierarchical, collective, and neutral), many young Japanese are egalitarian, individual, and emotional. One site would not fit all, but morphing could capture differences within the country. Field research is underway in Japan to determine the value of cultural morphing.

Trust is global, and many firms are formulating global Internet strategies. The usual approaches are to have one site for the world with only translation differences, or building separate sites for each country. We are planning to do research on how we can build the best global backbone and then morph it for cognitive and cultural individual differences between and within countries. With the growing globalization of Internet strategies, this type of research may provide the architecture for global trust and customer advocacy.

4.5 Future of Customer Advocacy

The Internet is continuing to evolve and Web 2.0 is the current term used to designate this new environment. Web 2.0 is characterized by user control and ownership of data

Figure 4.9 Suruga Bank Morphing (Reproduced by permission of Glen Urban/MIT)

collaborative work and social networks that have evolved out of communities. We know from past research that communities can increase trust (Urban, 2005), but we need to understand the functioning of these new social networks and their impact on purchasing decisions. New software makes Web 2.0 sites even more graphically attractive and easier to use. While some of the Web 2.0 features are evolutionary, rather revolutionary phenomena have taken place in Myspace and Facebook, Second Life, blogging, and customer complaint sites. Clearly, customer advocacy is a Web 2.0 concept, and as the Internet evolves, trust and advocacy will become more important and evident.

Firms that manifest trust by empowering customers with advice and transparency will gain sales and profit rewards. First movers will gain a preferred position in the customer's list of trusted suppliers, and later entrants will suffer. Can you imagine how tough it would be to displace your competitor if they had earned the most trust position in customers' minds? Now is the time for innovative firms to develop and implement trust generating customer advocacy systems.

References

Bart, Y., Shankar, V., Sultan, F., and Urban, G.L. (2005). Are the Drivers and Role of Online Trust the Same for All Web Sites and Consumers? A Large-Scale Exploratory Empirical Study. Journal of Marketing, 69(4), 133–152.

Fogg, B.J., Sohoo, C., Danielson, D.R., Marable, L., Stanford, J., and Tauber, R. (2003). How do users evaluate the credibility of web sites? A study with over 2,500 participants. In Proceedings of the 2003 Conference on Designing for User Experiences, pp. 1–15. San Francisco, CA.

Hofstede, G. (1980). Culture's Consequences: International Differences in Work-Related Values. Sage, Beverly Hills, CA.

Rousseau, D.M., Sitkin, S.B., Burt, R.S., and Camerer, C. (1998). Not So Different After All: A Cross-Discipline View of Trust. Academy of Management Review, 23(3), 393–404.

Saint Paul Insurance Co. (2000). Advertising Files. Saint Paul, MN.

Shankar, V., Urban, G., and Sultan, F. (2002). Online Trust: a Stakeholder Perspective, Concepts, Implications, and Future Directions. Strategic Information Systems, 11, 325–344.

Urban, G.L. (2005). Don't Just Relate-Advocate. Pearson Education and Wharton School Publishing, Upper Saddle River, NJ.

Part Two

Building the Infrastructure

5

The Semantic Web – from Vision to Reality

Paul Warren and John Davies

5.1 Setting Out the Vision

Since its conception in the early 1990s, by Sir Tim Berners-Lee, the Web has astonished us all, with its popularity and obvious value, both to the individual and the enterprise. Yet, even during the Web's first decade, Sir Tim himself was expressing dissatisfaction with it, and pointing the way towards a bolder vision.

The Web was conceived to make information held on computers available to people. Originally the people were researchers at CERN, the European particle physics establishment. Today, the users of the Web range from scientists looking for scientific knowledge to ordinary citizens looking for recipes or catching up with lost friends. People are highly flexible in their ability to interpret information. To take an example, imagine you are looking for a new car and are comparing prices on the Web. As you surf from web page to web page, the information will be presented in quite different ways, and the prices of the various cars will be located in different places on the various pages. Moreover, the real price will be different from the price as it first appears. You will have to take account of what extra features you want. Some manufacturers will include air conditioning as a basic feature; others will charge an additional payment. There may be discounts to take into account.

For humans, this is no problem – or at least not an insuperable one. However, if you want to automate this process, e.g. by using a software agent to do the job for you, then there are real difficulties. The problem stems from the initial choice of hypertext mark-up language (HTML) as the language of the Web. HTML permits the use of tags to describe the format of a web page, e.g. to indicate a bulleted list or to indicate that a line is a title or subtitle. However, the tags in HTML say nothing about the semantics, either of a web page itself or of the contents of the page. For the former,

ICT Futures: Delivering Pervasive, Real-time and Secure Services
Edited by Paul Warren, John Davies and David Brown
© 2008 John Wiley & Sons, Ltd

we might wish to indicate when the page was created, by whom, what topics it describes, etc. For the latter, we might want to indicate that a given text string represents a price, that it is in pounds or euros, that it is a basic price without extras, etc. In fact, there are a whole multitude of things we might want to say about the *semantics* of a page and its contents. It is this semantic information that a computer needs to interpret the page.

Moreover, to be interpretable by a computer, a web page needs to be free of any semantic ambiguity. For example, even if we know a price is quoted in pounds, we cannot be absolutely sure that we are talking about pounds sterling. There are, to the authors' knowledge, at least five other currencies called pound (e.g. Cypriot pound). This is exactly the kind of ambiguity which people resolve relatively easily, often from context, but which is difficult for computers to resolve.

Towards the end of the last decade, Berners-Lee and others were thinking about how to overcome these problems and create a truly *Semantic Web*. Early this century, they set out their vision of such a Semantic Web (Berners-Lee *et al.* 2001). In this vision, they described scenarios of what could be achieved with the Semantic Web. In one scenario, an individual identifies a doctor for medical treatment, taking into account factors such as the doctor's availability, geographical location, and whether a particular doctor in a particular hospital was included under a particular insurance plan. The article gave some pointers to how the Semantic Web would evolve, but was by no means a detailed blueprint. It was rather a manifesto, pointing the way towards a programme of work which has occupied this decade and will probably continue until the decade's end.

5.2 Describing Semantics

The challenge is to associate semantics with web pages and the content of web pages. The word *associate* is used deliberately. When using HTML, the formatting tags are embedded in the web page. With the approach to be described here, the semantic information may be in a separate file, but associated with the web page.

In any case, the semantic information can be interpreted by a computer. Again, the word *interpreted* is used deliberately, in place of *understood*, which is sometimes more colloquially used. We do not claim that computers will understand, in the sense in which humans understand. Rather, the semantics are made explicit. To do this, we draw on the field of knowledge representation, which has been studied since the early days of computer science. A language, resource description framework (RDF), has been defined by the World Wide Web Consortium (W3C) 'for representing information about resources in the World Wide Web' (Manola and Miller 2004). RDF is very simple. Each statement is a triple consisting of subject, predicate, and object. The subject and predicate are uniform resource identifiers (URIs)[1] and it is through the use of URIs that uniqueness is achieved. The object can be either a URI or a literal. Literals can be either plain, i.e. any text string in a natural language, or typed, e.g. numbers or dates.

---

[1] A URI can be regarded as a generalisation of a URL.

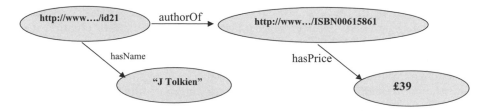

Figure 5.1   Linking RDF triples to form a graph

RDF triples can be linked to form a graph. Figure 5.1 illustrates such a simple structure which contains three triples. At the top, there is a triple which states that 'http://www . . . /id21' is the author of 'http://www . . . /ISBN00615861'. A human knows at once that the subject represents a person, and the object a publication. However, a computer will only be aware of the authorOf relationship. In practice, we might include the name (J. Tolkien) in the subject's URI, e.g. 'http://www . . . /person_J_Tolkien' rather than the anonymous 'http://www . . . /id21'. This would convey information to the human. However, a computer requires the explicit hasName predicate, and this is supplied by the triple on the left. On the right, there is a triple indicating the price of the book, represented by a typed literal. In this way, we create a graph structure, which is serialized using Extensible Markup Language (XML).

RDF lacks any predefined concepts, and this lack is addressed by RDFS. The acronym stands for RDF Schema. However, this is something of a misnomer. RDFS is not a schema language and, for example, does not bear the same relationship to RDF as XML Schema bears to XML. What RDFS does do is to introduce externally defined concepts such as Class, subPropertyOf, subClassOf, domain, and range.

This, in turn, enables the creation of an ontology, a word borrowed by computer scientists from philosophy. In computer science, the most frequently quoted definition of an ontology is by Gruber (1993): 'an ontology is a formal specification of a conceptualization'. In more colloquial language, an ontology is a formal model of a domain of knowledge, e.g. geography. Because it is formal, a computer can reason about it. For example, the computer can reason that, given that Microsoft is headquartered in Seattle, and Seattle is in the USA, then Microsoft is headquartered in the USA. This may seem trivial. However, imagine that we are searching a knowledge base for companies headquartered in the USA, and we have the geographical information in different forms, e.g. city, county, state.

RDFS is a very simple ontology language, and recently, the W3C has standardised a more sophisticated language known as Web Ontology Language (OWL) (see McGuinness and van Harmelen 2004). In fact, OWL comes in a number of different species with increasing richness of expression. With this increasing richness comes less computational tractability, and it is likely that for many applications, a simple variant of the language will suffice.

We can see that our Semantic Web makes use of a stack of languages: XML, RDF, RDFS, and OWL. In fact, Berners-Lee (2000) originally envisaged a 'layer cake' approach to the Semantic Web. We return to this in Section 5.6, where Figure 5.4 illustrates the layer cake. For a good introduction to the technical basics of the Semantic Web, including OWL, see Antoniou and van Harmelen (2004).

The development of the Semantic Web depends upon the existence of semantic metadata. There are two complementary approaches to metadata creation. Firstly, easy to use tools are available. This is crucial, since otherwise, there will be a natural reluctance to invest in the overhead of metadata creation. One example of such a tool is the semantic wiki discussed in the next section.

The other approach is to create the metadata automatically, i.e. by an analysis of the text. Here, there are two broad sets of techniques, which can be used together. Firstly, statistical and machine learning techniques[2] can be applied to a block of text, e.g. a web page, to create descriptive metadata. For an overview of such techniques see Groblenik and Mladenic (2006). The other set of techniques, drawn from natural language processing, relies on a detailed analysis of the grammatical structure of the sentences. With such techniques, it is possible to infer to a reasonable probability that certain text strings are the names of people, others those of companies. This is based on the format of the strings and the way they are used. As another example, we might have a web page, containing several paragraphs, with, at various points, the inclusion of 'Mr. Bush', 'The President', 'he'. Again, it is possible to infer that all three terms refer to the same person. For more detail on the use of human language technology for semantic annotation, see Bontcheva *et al.* (2006a).

The two approaches of manual metadata creation and automatic annotation may be combined in a semi-automatic approach. Here, the first pass is undertaken automatically, with a review allowing for human intervention to correct any mistakes. These corrections form, of course, the basis for learning to improve the results of subsequent automatic annotation. For a sample of the kinds of tools being developed, see Möller *et al.* (2006).

5.3 Early Applications

The last section described the technical fundamentals underpinning the Semantic Web. We now consider how the technology is being used.

A very early application of RDF was RSS. The acronym has a number of expansions (Really Simple Syndication, RDF Site Summary, Rich Site Summary), and has gone through a number of standards. The objective has remained the same: to be a format for frequently changing web sites, e.g. news sites. On the client side, a news aggregator is then used to display changing information from the user's favourite sites. When specifying the feeds, search terms can be used to indicate the user's precise areas of interest. All major newspapers and news providers now provide RSS feeds, e.g. *The Times*[3] and the BBC.[4]

Another early application of RDF which achieved rapid take-up has been created by the Friend of a Friend (FOAF) project,[5] whose objectives and basics are explained

---

[2] Statistical and machine learning techniques are both applied to aggregated text. Broadly speaking, the first term is used where there is a strong theoretical mathematical underpinning to the technique. The latter term is applied to more heuristic techniques.

[3] http://www.timesonline.co.uk/tol/audio_video/rss/.

[4] http://news.bbc.co.uk/1/hi/help/3223484.stm.

[5] http://www.foaf-project.org/.

by Dumbill (2002). FOAF is, in fact, an early example of social networking which enables individuals to describe themselves, their interests, and their relationships, thereby creating communities on the Web. Social networking is described in more detail in Chapter 2.

A more recent development is the emergence of the semantic wiki combining Semantic Web technology with the concept of the wiki. There are a number of implementations of this; one such, Semantic MediaWiki (Völkel *et al.* 2006), builds on the technology used by Wikipedia. An ordinary wiki makes extensive use of links between pages. The key extension of the semantic wiki is to enable the user to provide semantic information with the link. For example, the link from a page describing London to a page describing England can carry the semantic information 'is located in' or 'is capital of'. In addition, attributes, e.g. numbers, dates, and coordinates, can be included. The use of a very specific syntax for these links and attributes means that they are machine interpretable. This, in turn, means that the user can be provided with a semantic search capability; for example, he or she can search for the capital of England or ask to list all the cities located in England. These searches are semantic rather than textual because they do not rely on an accident of juxtaposition of text, which in any case can be misleading (e.g. 'Newmarket is the capital of England's horse racing industry'). Apart from being able to read and search the information in a Semantic MediaWiki, it can be made available as RDF to other applications.

A major feature of the semantic wiki is that there is no restriction on how the semantic links are annotated, i.e. there is no pre-defined ontology. One user might create a link between London and England annotated with 'is capital of', whilst another user might create a link between Paris and France annotated with 'capital city of'. This informal approach, borrowed from the world of folksonomies discussed in Chapter 2, is crucial to encouraging the creation of these links. It addresses the problem raised in Section 5.2, that without simplicity of approach, users will not create semantic metadata. Of course, users are encouraged to reuse semantic terms rather than to invent their own. Users can also be encouraged to create equivalences between links where appropriate and also to indicate where one link type is a special case of another, e.g. 'is capital of' is a special case of 'is located in'. The semantic wiki is an example of the fusion of the methods of the Semantic Web, with its formally defined ontologies, and the Web 2.0 approach, with informally defined folksonomies. We are likely to see more of such examples in the future.

Apart from these general applications, there are also some specific domains where significant ontologies already exist. One such is the health sector, where the nature of the phenomena being described gives rise to very large ontologies (Wroe 2006). Another domain where significant ontologies have been developed is that of geographical information systems, e.g. see Fonseca *et al.* (2002).

5.4 Semantic Web Services

Most of what has been said so far has been about finding and using information on the Web. However, the Web also contains computational objects, i.e. executable code, which can undertake tasks. Typically today, these computational objects are made available as web services, i.e. they are available in a registry employing the Universal Description, Discovery and Integration (UDDI)[6] protocol, and they are accessed using SOAP[7] (Mitra and Lafon 2007). The problem is that the descriptions of web services in a UDDI registry are written to be interpreted by people, not computers. In order to realise the Semantic Web vision, web services need to be semantically annotated so that they can both be easily found and then automatically combined to form more complex ones. We envisage a situation where we search for a web service to undertake a particular function. One may already exist for this function. Alternatively, we may be presented with a new web service automatically composed of a number of simpler ones.

What is required is an ontology for describing web services. One such is the Web Services Modelling Ontology (WSMO) (Roman *et al.* 2006). Once web service has been described semantically, we can use a Semantic Web service browser to locate a web service. Because, using WSMO, the inputs and outputs from each web service are described semantically, it is possible to automatically compose new web services from existing ones. Finally, from the browser, we can invoke the new web service. Ideally, such a browser could be used by nonprogrammers.

Davies *et al.* (2004) present a vision of web services and describe a Semantic Web service browser, reproduced here as Figure 5.2. The left-hand pane shows the categories in the domain ontology. In the figure, the user has selected SMS. The right-hand pane shows the services which were matched to this category. The user is then free, for example, to invoke the service or add the service to the pool for combining with other services.

Duke *et al.* (2005) describe an early application of WSMO to create a business-to-business architecture, specifically in the telecommunications industry. The architecture supports interaction between a telecommunications operator and broadband service providers. The use of WSMO enables 'the semantics to be explicitly represented in a standard ontology based on industry standards, rather than being tied up within the message format'. Such an approach involves an initial overhead for the telecommunications operator in using WSMO to represent its interface. This done, service providers will be able to integrate more quickly, with fewer errors, and requiring less support.

The next chapter will describe an industry-wide move towards a service oriented architecture. The chief element of this is the use of loosely coupled components, frequently web services, which interact through well-defined interfaces, thereby avoiding the need for systems integrators to have any understanding of the internal implementation of those components. Semantics will be a key enabler for this vision. In a world with tens of thousands, or more, of such components distributed across the Web, then only through semantics will we be able to locate the services we need and to combine them to create the end service we want.

---

[6] http://www.uddi.org/.
[7] The SOAP acronym was originally expanded as the Simple Object Access Protocol and is now sometimes also expanded as the Service Oriented Architecture Protocol. In any case, the unexpanded acronym is commonly used.

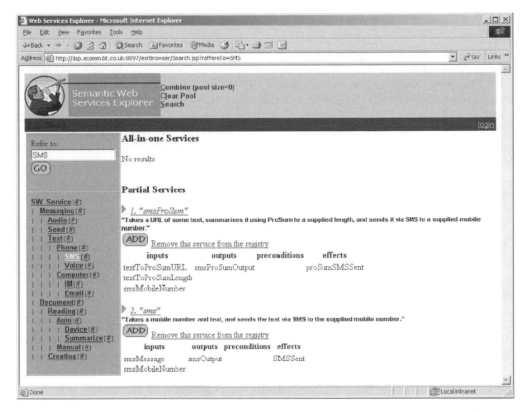

Figure 5.2 A Semantic Web service browser (Reproduced by permission of BT)

5.5 Semantic Web Technology in the Enterprise

Many early applications of Semantic Web technology have been within the enterprise, where the problems of scale are less challenging than on a global Semantic Web. Here, Semantic Web technology can be used to address two major challenges: that of semantic incompatibilities between corporate databases and the challenge of tapping into the knowledge held in unstructured information.

5.5.1 Semantic Data Integration

Organisations are bedevilled by semantic incompatibilities. Typically, an organisation will have tens of databases, with incompatible schemas. On the one hand, the same term will be used differently in different schemas; on the other hand, different terms will be used to represent the same concept. This situation occurs often enough within single organisations. It occurs all the more when mergers and acquisitions form new conglomerate organisations. It occurs also when organisations work together in a supply chain and semantic incompatibilities arise through differences in terminology between the organisations. The significance of overcoming semantic incompatibilities has been stressed by McComb (2004) in a book which provides a good introduction to the business issues of using semantic technology.

One solution, frequently implemented today, is to create bilateral mappings between each pair of databases. The defect of this approach is obvious. The number of mappings grows rapidly with the number of database. An alternative approach is to use a central broker with mappings between each database and the broker. An overarching ontology is created in the broker, and then mappings are created between each database schema and the ontology. Using this approach, there will be one bilateral mapping for each database. Of course, this broker architecture is well known in computer science and has been used for decades to overcome syntactic incompatibilities, i.e. the use of different formats or protocols. The difference here is that the mappings are at the semantic, rather than at the purely syntactic level.

The mappings do first need to be created. This is usually done using a graphical tool. One such, from ontoprise GmbH[8], is shown in Figure 5.3. Drag-and-drop functionality is used to create mappings, whilst at the same time, the system undertakes consistency checks. The left- and right-hand sides show portions of two different ontologies, and the mappings are represented by lines between them.

The ideal is to create the mappings automatically, and this is indeed the subject of current research. The reality is that total automation is highly unlikely ever to be possible. However, a semi-automatic approach is possible, where mappings are proposed, and a user can accept, reject, or edit them.

This approach is applicable to information stored in structured form, in particular, in relational databases with well-defined schemas. However, an increasing quantity of information is unstructured, e.g. emails, text documents, intranet pages; or alternatively is semi-structured, e.g. in spreadsheets where tables are used but there are unlikely to be well-defined schemas. Forrester Research, a technology and market research company, estimates that 80% of company information is unstructured (Moore 2007). The need is to mine this unstructured information and to merge it with the structured.

5.5.2 Knowledge Management and Business Intelligence

The challenge of unstructured information is to exploit the information based on the underlying concepts, rather than the particular textual expression of those concepts. Take search as an example. To date, search technology, although impressive, is based chiefly on looking for coincidences of text strings. A search for 'engine' will not usually find occurrences of 'motor', although they both represent the same concept. A starting point to overcome this is the use of a system such as Wordnet,[9] which organises words into synonym sets, so that a search for 'engine' can be expanded to look for 'motor'. The next stage would be to include the ability to reason; an example of the use of reasoning about geographical information expressed in an ontology has been given in

---

[8] http://www.ontoprise.com.
[9] http://wordnet.princeton.edu/.

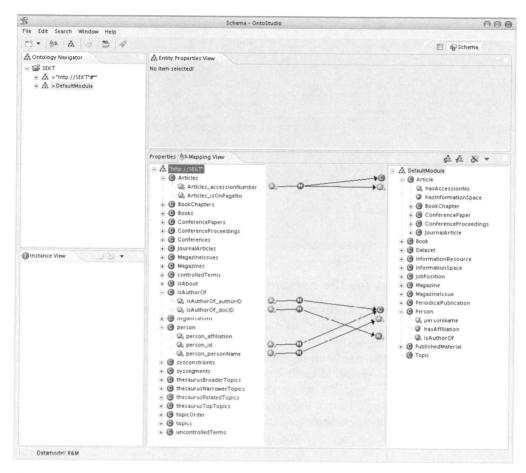

Figure 5.3 A graphical editor for ontology mappings. (Reproduced by permission of ontoprise GmbH)

Section 5.2. A further stage would be to use the techniques of natural language processing also described in Section 5.2, so that references to an individual can be found when that individual is referred to simply as 'he' or 'she'.

One example of semantic search, SEKTagent, developed in the SEKT project,[10] is described in Bontcheva *et al.* (2006b). An interesting example of how an enterprise can use the same technology to scan the Web for business intelligence, specifically in this case in the recruitment industry, is given in Kiryakov (2006).

Extending such techniques beyond search enables an organisation's unstructured information to be mined and merged with information in its structured databases. For a more detailed overview of how semantic technologies are being used in the organisation to manage knowledge, see Warren (2006).

---

[10] http://www.keapro.net/sekt/.

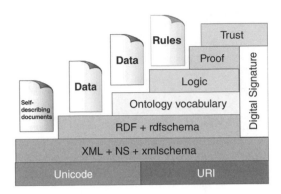

Figure 5.4   Semantic Web layer cake. (NS, namespace)

5.6 The Next Steps for the Semantic Web

We are still some way from achieving the original vision of the Semantic Web. For a review of the remaining challenges, to complement the discussion in this section, see Shadbolt *et al.* (2006). Nonetheless, a number of significant applications are now available. Semantic wikipedia has been described in Section 5.3, where its importance as a bridge between formal semantic techniques and the informal techniques of Web 2.0 has been stressed. For a wider view of what is possible, the reader is referred to the results of recent Semantic Web Challenges. These are annual competitions to create practical Semantic Web applications. Shortlisted papers are presented at the International Semantic Web Conference,[11] and the best are published in the *Journal of Web Semantics.*

To compare progress against the vision, we consider the so-called 'layer cake', shown in Figure 5.4. This is the technology stack which underpins the Semantic Web (Berners-Lee 2000).

Section 5.2 has already described many of these layers. At the bottom layer, we use URIs to ensure uniqueness, plus Unicode[12] to provide a unique encoding capable of representing all human languages. Above this is the use of XML, XML Schema and XML namespaces. Above this again is the use of RDF and RDFS for basic ontology modelling. The ontology vocabulary layer has now been filled by the use of the OWL language to describe richer ontologies. At the logic layer, languages now exist to define rules, e.g. business rules or IT management policies, and there are now many efficient reasoners available.

This leaves two top layers where much is yet to be done: proof and trust. The need for both was explained in the original visionary article (Berners-Lee *et al.* 2001). We need not just to have the correct answer, but to have an auditable proof that it is correct. In business, for example, disputes about payment may arise. So proof is now

---

[11] e.g. see http://iswc2006.semanticweb.org/submissions/web_challenge.php.
[12] http://www.unicode.org.

an active area of research within the Semantic Web community. Above this in the layer cake, but very much related to it, is trust. This requires an understanding of provenance, i.e. how can we be sure of the information's source. A discussion of both proof and trust is given by da Silva *et al.* (2003). At the 15th International World Wide Web Conference[13] in 2006, there was a workshop devoted to 'Models of Trust for the Web'.

Finally, the Semantic Web may at first be more like a set of semantic islands. These will interconnect syntactically through the basic Web, but each having their own ontologies. Of course, wherever possible, people should share ontologies. Still, different ontologies will spring up, to suit different conditions or by accident. So, to encourage interoperability, we shall need globally scaleable mappings between ontologies.

5.7 Conclusions

This chapter has described the original vision of the Semantic Web, the technologies needed to achieve that vision, some of the initial applications, and the chief remaining challenges. Whilst describing what needs to be done to achieve the original vision, the chapter has demonstrated that there are already applications which can be used to achieve real benefits for individuals and enterprises.

References

Antoniou, G. and van Harmelen, F. 2004. A Semantic Web Primer, MIT Press, Cambridge, MA.

Berners-Lee, T. 2000. Semantic Web on XML, *XML 2000*, http://www.w3.org/2000/Talks/1206-xml2k-tbl/slide1–0.html

Berners-Lee, T., Hendler, J., and Lassila, O. 2001. The Semantic Web, *Scientific American*, May 2001.

Bontcheva, K., Cunningham, H., Kiryakov, A., and Tablan, V. 2006a. Semantic Annotation and Human Language Technology, *in* Davies *et al.*, eds. *Semantic Web Technologies*, pp. 29–50, John Wiley & Sons, Ltd.

Bontcheva, K., Davies, J., Duke, A., Glover, T., Kings, N., and Thurlow, I. 2006b. Semantic Information Access, *in* Davies *et al.*, eds. *Semantic Web Technologies*, pp. 139–169, John Wiley & Sons, Ltd.

da Silva, P., McGuinness, D., and McCool, R. 2003. Knowledge Provenance Infrastructure, IEEE Data Engineering Bulletin 26, 4, 26–32.

Davies, N., Fensel, D., and Richardson, M. 2004. The Future of Web Services, *BT Technology Journal*, 22, 1, 118–130.

Duke, A., Davies, J., Richardson, M. 2005. Enabling a scalable service-oriented architecture with semantic Web Services, *BT Technology Journal*, 23, 3, 191–201.

Dumbill, E. 2002. XML Watch: Finding friends with XML and RDF, http://www.ibm.com/developerworks/xml/library/x-foaf.html

Fonseca, F.T., Egenhofer, M.J., Agouris, P., and Câmara, G. 2002. Using Ontologies for Integrated Geographic Information Systems, *Transactions in GIS* 6, 3, 231–257.

Grobelnik, M. and Mladenic, D. 2006. Knowledge Discovery for Ontology Construction, *in* Davies *et al.*, eds. *Semantic Web Technologies*, pp. 9–27, John Wiley & Sons, Ltd.

Gruber, T. 1993. A Translation Approach to Portable Ontology Specifications, Knowledge Acquisition 5, 2, 199–220. http://ksl-web.stanford.edu/KSL_Abstracts/KSL-92-71.html

Kiryakov, A. 2006. Scalable Semantic Web Technology for Recruitment Intelligence. Presentation at the Business Applications session at WWW2006. http://www2006.org/speakers/kiryakov/

---

[13] http://www2006.org/.

Manola, F. and Miller, E. 2004. RDF Primer, http://www.w3.org/TR/rdf-primer/

McComb, D. 2004. *Semantics in Business Systems*, Elsevier, San Francisco, CA, (see Chapter 12 for a discussion of integration).

McGuinness, D. and van Harmelen, F. 2004. OWL Web Language Overview, http://www.w3.org/TR/owl-features/

Mitra, N. and Lafon, Y. 2007. SOAP Version 1.2 Part 0: Primer (Second Edition), http://www.w3.org/TR/2007/REC-soap12-part0-20070427/

Möller, K., de Waard, A., Cayzer, S., Koivunen, M., Sintek, M., and Handschuh, S. (editors) 2006. Proceedings of the 1st Semantic Authoring and Annotation Workshop, http://sunsite.informatik.rwth-aachen.de/Publications/CEUR-WS/Vol-209/

Moore, C. 2007. Forrester's Connie Moore: The World of Content Management is Changing, Information Indepth, Oracle, http://www.oracle.com/newsletters/information-insight/content-management/feb-07/forrester-future.html

Roman, D., de Bruin, J., Mocan, A., Toma, I., Lausen, H., Kopecky, J., Bussler, C., Fensel, D., Domingue, J., Galizia, S., and Cabral, L. 2006. Semantic Web Services – Approaches and Perspectives, *in* Davies *et al.*, eds. *Semantic Web Technologies*, pp. 191–236, John Wiley & Sons, Ltd.

Shadbolt, N., Hall, W., and Berners-Lee, T. 2006. The Semantic Web Revisited, *IEEE Intelligent Systems*, May/June 2006, 96–101.

Völkel, M., Krötsch, M., Vrandecic, D., Haller, H., and Studer, R. 2006. Semantic Wikipedia, *International World Wide Web Conference WWW2006* http://www.aifb.uni-karlsruhe.de/WBS/hha/papers/SemanticWikipedia.pdf

Warren, P. 2006. Semantic Information Management – Managing Information Through Meaning Not Words, *Journal of The Communications Network*, 5 Part 4, October–December 2006, 58–68.

Wroe, C. 2006. Is Semantic Web Technology Ready for Healthcare? 3rd European Semantic Web Conference, June 2006.

6

Flexible ICT Infrastructure

Mike Fisher and Paul McKee

6.1 Introduction

Information and communication technologies are changing the structure of the global economy, allowing people and businesses to interact wherever they are in the world. New opportunities and competition can appear very quickly, and businesses need to be able to respond. The ICT systems deployed in most established organisations of any size are not well suited to a rapidly changing business environment, and a new approach is required. Ubiquitous network connectivity is changing the nature of computer applications. Software is increasingly distributed across a number of computers, and different components execute concurrently. Building well-engineered applications in this environment is difficult. There are some simplifying ideas – viewpoints and distribution transparencies that should be able to help manage some of this complexity.

In this chapter, we focus on the physical infrastructure of computers and networks that underpin all ICT applications. This is typically a diverse mixture of different kinds of hardware from different vendors. Recently, a number of technologies loosely referred to as 'virtualisation' have started to emerge. These provide low-level flexibility that can support the construction of agile and responsive ICT systems. An architectural approach to building applications based on loosely coupled services is becoming widely accepted. The same approach fits very well with virtualised ICT resources and allows heterogeneous components to be viewed as a coherent infrastructure. This is still at a relatively early stage. The overall economic utility is expected to increase rapidly with the size of the interconnected infrastructure – in a similar way to the growth of the Internet or the Web. While some of the early steps towards large scale service oriented infrastructure (SOI) can be foreseen with reasonable confidence, there remain some major research challenges as they approach global scale.

ICT Futures: Delivering Pervasive, Real-time and Secure Services
Edited by Paul Warren, John Davies and David Brown
© 2008 John Wiley & Sons, Ltd

6.2 ICT in the Global Economy

ICT has already had a profound impact on the global economy, and widespread recognition of this has led to significant investment in ongoing development and the
potential for even greater impact. The growth of ICT has accelerated the creation of
international linkages and interdependencies between companies. The most successful
companies in the new global economy are agile and are always looking for opportunities to reduce their costs. The improved communication links provided by ICT and
the increasing commoditisation and specialisation of parts of the value chain allow
the production process to be broken into its component parts and for activities to
be performed in the lowest cost geographical location. Consumers and businesses
are thus able to take advantage of lower prices resulting from increasing online
competition.

Governments increasingly recognise the importance of ICT to their respective economies. A recent report prepared by the President's Council of Advisors on Science
and Technology (PCAST 2007) reports work by Jorgenson and Wesner that states that
networking and information technologies have accounted for 25% of economic growth
since 1995 in the USA whilst representing only 3% of the GDP. The report states that
leadership in the field of networking and information technology is essential to the
nation's global competitiveness and economic prosperity. In his foreword to the European e-Business Report , e-Business Watch (2007), the Vice President of the European
Commission states that the ICT industry is a major contributor to growth with an
annual average growth rate of 6% until 2008 and that as companies increasingly use
ICT for conducting business, e-Business has become a critical factor for competitiveness and productivity growth.

Governments in those countries with significant Internet penetration increasingly
use ICT to deliver more services to their citizens with reduced costs. End users are
also turning to ICT to deliver entertainment in ways that allow a level of personalisation never before achievable. They are able to select not only which programmes to
watch but also when to watch them, enabling a 'pick and mix' approach to entertainment that may threaten conventional broadcast television scheduling. National barriers are being broken down at the personal level by global social networking sites such
as Facebook and MySpace.

In the work arena, teams may now be formed that cross national, company and
cultural barriers to make use of the skills and expertise of individuals. This interaction
is facilitated by ubiquitous network connectivity and the wide range of collaboration
and communication tools available today. The flexibility this provides drives innovation. Access to the global networked economy is seen as an important prerequisite for
development in many of the world's poorer countries. Against this background, it
seems that ICT is of great importance to the global economy, and that importance in
unlikely to reduce in the foreseeable future.

6.3 ICT Resources

The ICT resources used in any enterprise of more than moderate size are highly
complex and consist of a wide range of computing and network hardware, middleware

products and software. These resources, together with their associated management systems, are becoming increasingly important to the success of the enterprise. ICT systems, such as a billing system or a payroll system, are typically built to address a single set of goals, specified in terms of a clearly defined set of requirements and deployed on a dedicated set of computing resources, with little attention paid to possible changes that are not anticipated from the outset. While this makes sense in terms of providing a cost-effective solution, such systems are difficult and expensive to modify. This may limit the ability of the enterprise to adapt its processes to respond to changing conditions. ICT becomes an inhibitor rather than an enabler of flexibility. Extending beyond the boundaries of a single enterprise presents additional challenges, and a new, more flexible approach is required that allows disparate ICT resources to be managed in a consistent way as a general-purpose infrastructure, within a single enterprise or as part of a ubiquitous infrastructure for applications in the global economy.

There are several different kinds of ICT resource that need to be incorporated into a general-purpose infrastructure:

- Computer processing provides the ability to execute software. A cluster of physical computers may be combined as a single computing resource or, alternatively, a single physical computer may be divided into a number of individual computing resources.
- Storage provides the ability to maintain persistent data and to retrieve it as and when required. There are several distinct kinds of storage system in use, with different levels of performance, capacity and cost. Direct attached storage has a storage device essentially as part of a host system, the simplest example being a server with an internal hard drive. Network attached storage (NAS) offers greater scalability and flexibility by separating the storage from the servers. It provides file-based storage to a number of servers from a specialised storage device over the same internet protocol (IP) network used to interconnect the servers. Storage area networks (SANs) offer block-level storage which allows more direct access to storage volumes. This allows an application, such as a database, to have more control over its data structures than it would have with NAS. Higher performance can be achieved in this way. However, the application may be more complex as it has to provide for itself any file system functionality it needs. SANs generally use a special dedicated network to completely separate data storage and data processing. This separation and the use of optimised protocols give high performance and reliability.
- Generic software such as operating systems, middleware, application servers and other software component containers can also be regarded as ICT resources. These form part of the environment for which a piece of application software can be developed.
- Network connectivity allows different resources to be connected together. It also enables applications to be distributed across different physical hardware and to interact with users and other applications. Connectivity resources include local area networks within a data centre, specialised networks to support data transfer in storage systems and wide area networks for communication with systems potentially anywhere in the world.

Within each category, there is considerable variation in terms of different technologies and products, often with very particular target application areas, strengths and limitations. An essential aspect of moving towards an infrastructural view of ICT resources is to identify the right levels of abstraction. This is primarily so that users or application developers can benefit from the capabilities of the infrastructure without having to deal with (or even know about) the complexity of the systems that make up the infrastructure. An infrastructure provider clearly needs a different level of abstraction since awareness of the details of particular pieces of equipment and their configuration is essential for effective management. In fact, an infrastructure provider may need multiple views, since it must deal both with customer relationships and low-level technical configuration.

6.4 Distribution Transparencies – Hiding Complexity

Conventional software applications are generally written with a particular hardware environment in mind, for instance, a PC running Windows, a workstation running a version of UNIX or Linux, a games console, a mobile phone or other specialised systems. This presents the application developer with a set of well-understood programming interfaces, constraints and behaviours. Software can be optimised based on detailed knowledge of the target environment, either directly by the developer or supported by machine-specific libraries and tools such as compilers. Communication between different software modules is generally local to a single machine. Failure modes can be predicted and are usually all-or-nothing. Testing is also relatively straightforward as the deployment environment can be accurately replicated during development, and automated tools can achieve reasonable coverage of different aspects of the application's behaviour.

Networked applications have significantly different characteristics, and this makes the task of the application developer considerably more complex. It can no longer be assumed that the hardware environment is completely known at development time. Since networked applications are inherently distributed and software modules may be on different machines, details of the communication links can become very important to the overall behaviour. Partial failures can occur; communication delays can vary over wide ranges including disconnection or partitioning. Comprehensive testing becomes much harder and more time-consuming, and the confidence that the application will perform as expected in practice is generally lower than for a conventional application. The specialised skills needed to develop high-quality networked applications present a barrier to creativity and mean that the full potential of ICT in the global economy is hard to realise.

Distributed systems are complex partly because there are many different people involved, playing very different roles. These include software developers, service providers and infrastructure providers. Their different roles and associated needs mean that they have very different views of the same system.

In particular, application development needs to be separated from deployment with clear limits to the developer's knowledge of and control over the run-time environment. Recognising that different viewpoints are required is a way of dealing with some of the complexity in distributed systems. International Standard 10746-2 (1996) describes the idea of distribution transparencies – properties of a system that hide some aspects of the system behaviour. This can insulate an application from changes in its supporting infrastructure and means that a developer can work with a simple model of the deployment environment – very like that of a single, predictable platform.

Many of the problems faced by networked applications arise simply from the fact that they are distributed, and distribution transparencies can be used to isolate and help solve them. There are several commonly considered transparencies. For example, failure transparency can hide the fact that individual components can fail and the fault tolerance mechanisms put in place to deal with these failures. The infrastructure simply behaves as if failures do not occur from the point of view of the application. Migration transparency can hide the movement of software modules or data between different machines, reducing communication latency or maintaining responsiveness by load balancing. Other commonly considered transparencies include access, persistence, security, transaction, location and replication.

Transparencies are generally associated with services provided by the infrastructure, and using these services can make many problems encountered in developing distributed applications tractable. The application developer (or user) does not need to know the details of the implementation or configuration. It is therefore worth providing solutions as part of the infrastructure. This means that their cost can be spread over many applications and users, justifying their optimisation by experts. A service-based approach to providing a platform for networked applications also means that the provider of the platform can make changes to the underlying infrastructure (hardware, network topology, run-time environment) without affecting the applications that use the services.

6.5 Different Views of Infrastructure

Distribution transparencies provided by infrastructure services have focused on making application development and deployment easier. This will stimulate innovation in the networked economy. However, there is a balance between ease of development and control of behaviour. This situation can clearly be seen in the Internet which offers a simple abstraction (IP) to wide area communication services. Taking this approach has led to its emergence as a ubiquitous, global infrastructure. However, the best-effort, unpredictable service levels available mean that it is not adequate for many business-critical applications. If the particular features of the supporting infrastructure are hidden, an application cannot adapt itself to them to obtain improved performance.

This indicates that there is a need to provide different views of the infrastructure to users with varying levels of technical awareness. Some developers or users will find a simple, general-purpose interface most useful. Others may want a more complex way of interacting that allows them more visibility of, or control over, the infrastructure. This makes their applications more complex but also potentially better performing.

A very flexible way of representing the capabilities of ICT infrastructure to its users is as a set of discrete services that can be selected in various combinations as required. Many independent providers will offer ICT infrastructure in the global economy, so the user may be faced with the challenge of combining services which are not coordinated by a single provider.

6.6 Resource Virtualisation

Virtualisation, when used in the computing industry, is a very broad term that refers to the abstraction of resources. This means the hiding of the physical characteristics of the resources and focusing on the way in which other resources, systems or end users interact with them. This includes making a single physical resource such as a server appear to function as multiple logical resources, or it can include making multiple physical resources appear as a single resource, an approach often used for storage devices. To illustrate the benefits that might be expected from adopting a virtualisation strategy, the rest of this section focuses on server virtualisation. There are a number of products and technologies that are all grouped under the umbrella of virtualisation and, in particular, server virtualisation. These are described in detail in manufacturers' literature and in articles such as that of Phelps and Dawson (2007).

At the time of writing, surveys suggest that the take up of server virtualisation techniques is already significant. According to Didio and Hamilton (2006), 76% of companies already use or plan to use server virtualisation. The benefits that users expect to obtain from virtualisation are in two areas, reduction in cost and increase in operational flexibility.

Once the direct relationship between an operating system or application and its supporting hardware is removed by the abstraction layer provided by virtualisation, sharing of hardware resources is much easier to achieve. The ability to have multiple operating systems and therefore multiple applications running on a single physical server allows a reduction in the overall number of servers being operated by an organisation. Typical server utilisation is low and can be increased by consolidation, with a number of significant economic benefits: fewer machines need to be purchased and maintained; the overall energy consumption can be reduced in line with the machine reduction, and space is freed up in data centres by the reduction in the number of physical systems.

Other less obvious benefits impact both costs and business agility. Once a choice of virtualisation technology has been made and adopted throughout an organisation's IT infrastructure, a number of other benefits become apparent. The costs of service deployment are reduced, as a virtual machine image can be deployed quickly and easily on any machine. The same image may be deployed multiple times on any number of servers to accommodate changes in demand, or as a low-cost fault tolerance or business continuity solution by moving images from failed to running machines. The abstraction layer introduced by virtualisation also reduces vendor lock in for the hardware, since a virtual image will execute on any hardware that supports the chosen virtualisation technology. This also reduces the number of virtual images that need to be maintained by the organisation.

Of course, the flexibility introduced by virtualisation is not without problems. The real server sprawl reduced by consolidation is replaced by virtual server sprawl, which may be harder to manage as it is easy to introduce new virtual machines into the system. The use of shared resources complicates performance monitoring, and the impact of system failures may be more complex to diagnose and react to. This has led to an increased interest in specialist fault tolerant hardware to lessen the impact of failure, but this of course impacts on the potential cost savings that might be obtained. Licensing is another important consideration in any decision to adopt server virtualisation. The main issue is commercial rather than technical. Software vendors generally define licence terms with the assumption that their software will be installed on a clearly identified and fairly static set of computers or individual users. If new approaches to licensing are not introduced, virtualisation and the flexibility it enables may require users to buy significant numbers of additional licences, again reducing potential cost savings. It is not yet clear what licensing models will be acceptable to both software vendors and their customers.

6.7 Architectural Principles

Service orientation is an architectural style that can be used for building distributed computing applications. It is based on the encapsulation of business functions as services that can be assembled in a loosely coupled fashion to provide applications quickly and cheaply. Service orientation is not a single product or set of standards. It is an approach for building agile and flexible business applications based on a number of defining architectural characteristics or principles. The current focus of service orientation is on business processes and on the flexibility that can be gained by separating interface, implementation and protocol binding in the ICT services that support them. This allows choices associated with the behaviour of the service to be made when it is deployed onto a particular set of resources, rather than as part of the implementation.

The construction of software systems based on the use of services has attracted much attention in recent years as service oriented architecture (SOA). OASIS (2006) gives a clear definition of a number of important concepts in its Reference Model for Service Oriented Architecture. The basic motivation behind SOA is to provide a way of 'organising and utilising distributed capabilities that may be under the control of different ownership domains'. This is the problem faced in matching the needs of users with the functionality offered by infrastructure providers.

A well-designed service must be modular and self-contained so that it is capable of composition with other services into a more complex application. It should have a level of granularity that allows it to operate at a business level – providing a clear and understandable function to its users. It should also encapsulate the data necessary to operate effectively. Services should be replaceable by the service provider to allow upgrades or bug fixes. User should also be able to replace a service used in one of their own business processes or workflows with an alternative, possibly from a different provider. In each case, there should be no unintended side effects.

The function of a service should be described by a precise agreement that sets out in detail the terms that define success criteria and the expectations of both the provider and consumer. Service implementations should give the service provider the flexibility to choose where to execute it so that the agreement can be fulfilled cost-effectively.

The agreement should specify both functional and nonfunctional aspects of the service delivery and should include such things as performance, availability, price and reliability. Typically, if the service crosses an organisational boundary, the agreement will be in the form of a carefully prepared service level agreement (SLA), although the SLA approach has advantages even within a single organisation. The agreement sets out the expectations and obligations of both parties involved and provides details of the relevant performance metrics and their method of measurement. The SLA is an important commercial tool that may be used as part of the service selection and advertising process. The agreement should not prescribe how the service is to be delivered. This maintains the freedom of the service provider to choose the most cost-efficient method of delivery without interference by the consumer as long as the terms of the SLA are met.

There are advantages in deploying service implementations that can support multiple users. In addition, the service code should be reusable in deployments for different users and organisations. These features confer advantages on both parties. Development costs are amortised across a number of users, reducing the overhead per consumer, and capacity can be more efficiently utilised by sharing.

At the implementation level, the code modules that make up a service should be packaged for deployment in an automated way on a range of hardware. Providers and consumers of services will be distributed across the network, and the use of standard protocols eases communication.

The provision of precise machine-readable service descriptions allows other services to automatically discover and dynamically build requests for the service, enhancing the flexibility of the system. The interface metadata has to be available independently of the executable code. This enables consumers to create messages of the right format and content to communicate with remote services in a loosely coupled way. In general, all of the data the service needs in order to function should be encapsulated so that the only way of interacting with the service is through its published interfaces.

6.8 SOI

Following the architectural principles described in the previous section, the software development world is becoming increasingly service oriented. Applications based on SOAs such as web services are also proliferating. In order to be able to provide the same level of flexibility and responsiveness to change in the infrastructure, there is a move towards SOI where networks, servers and storage hardware are made available as a set of services. SOI, in part enabled by the advances in resource virtualisation, supports the SOA and facilitates the reuse and dynamic allocation of hardware resources. The entire supporting infrastructure has the same service-based structure as the software that is executed on it, whether the resources are physical or virtualised. The business drive for this is the fact that users want a consistent end-to-end approach to support agile business processes, and this will require end-to-end management across the combined software and hardware infrastructure. End users will increasingly seek a single unified SLA that covers the end-to-end delivery of any business process, measured against a meaningful business metric. This standardisation of delivery should result in lower operational costs and savings that can be passed on to the service consumer.

6.9 Standardisation

Open standards are essential to achieve the full benefits of a service oriented approach to ICT infrastructure. Many different parties with different perspectives and concerns may need to cooperate in the provision and effective use of a large scale ICT infrastructure, particularly as interactions move beyond the boundaries of a single organisation. Common architectural and design patterns for ICT infrastructure and its management will certainly be important in achieving appropriate distribution transparencies for applications.

OASIS (2006) identifies three main aspects that are relevant to the usage of a service: visibility, interaction and effect. This provides a useful classification of areas where standard approaches are required. Visibility refers to the need for users and providers of a service to be aware of each other and to establish the context necessary for use of the service. This includes a precise description of what the service does, what the user can expect of it, and any obligations and constraints on both parties. This should be expressed in the language of the user – a point that emphasises the need for a range of services at different levels of technical awareness. This shared context corresponds to the SLA introduced above. Interaction refers to the need for the user to be able to call on the capabilities provided by the service and for any relevant information to be communicated between user and provider. It includes protocols for message exchange or invocation of actions, and can be regarded as defining the service interface. Interaction includes not only the mechanics of communication but also the interpretation of the information exchanged. The effect of using an infrastructure service might include returning some information or changing the configuration of the infrastructure. Effect relates only to changes in the state of the infrastructure that are relevant to the user. Many internal details of the service implementation will be hidden, and the effects of interest are those that change the user's view of the infrastructure – at the appropriate level of abstraction.

The current specification and standardisation landscape for SOA is very complex with hundreds of different activities addressing different aspects of building systems with a service-based approach. However, according to Heffner (2006), very few standards have actually been widely adopted. These really only provide basic visibility (e.g. UDDI, WSDL)[2] and interaction (XML, XML-Security, SOAP, WSDL), but a more coherent and consistent approach is necessary. Priorities for ICT infrastructure include standard approaches for description and exposure of infrastructure as services.

There are major efforts aimed at developing information models for ICT infrastructure, notably the Common Information Model (CIM) from the Distributed Management Task Force,[1] which has a strong IT system focus, and the Shared Information and Data Model (SID) from the Telemanagement Forum, which also has a strong emphasis on networks. CIM and SID each define a technology neutral management

---

[1] http://www.dmtf.org.
[2] UDDI: Universal Description, Discovery and Integration. WSDL: Web Services Description Language.

model of physical and logical devices, and there are efforts underway to harmonise the two approaches. They provide a comprehensive structure that can help an infrastructure owner with achieving a consistent approach to management, but they do not provide the appropriate abstractions for exposing infrastructure services to application developers and users as services in an accessible way.

6.10 Outstanding Challenges

There has been considerable activity recently in developing products and solutions for flexible ICT infrastructure. Significant challenges remain, particularly in building and managing large scale heterogeneous deployments.

Current virtualisation technologies are diverse, and it is not clear how they should be combined as components of a single infrastructure. A consistent way of presenting ICT resources to applications and developers is required. This needs to offer appropriate abstractions of the underlying hardware in an accessible way. Once this becomes established, we can expect better support from development tools. This will reduce a significant barrier to producing innovative applications.

Management of both applications and their supporting infrastructure is made more complex when the flexibility of ICT infrastructure is used, partly because configurations can change rapidly and partly because of the uncertainties inherent in many applications sharing the same resources. While the developer and user of an application do not know or care which resources are being used at any given time, the infrastructure provider needs to. If there is a hardware failure, it is important to know which applications are affected and how to mitigate the failure.

Without knowledge of the infrastructure (servers, storage devices, network links) being used to support their applications, customers can only purchase infrastructure services based on performance guarantees. This strongly suggests a need for SLAs that specify behaviour in a business relevant way, and means that the service provider has to understand the implications for configuration and management of infrastructure to support a large number of customers with different SLAs simultaneously. The additional complexity of management in this environment means that a high degree of automation is essential. Ensuring stable, robust and predictable behaviour in a large scale, shared infrastructures will be a major challenge.

6.11 Conclusion

In this chapter, we have described some current developments in ICT infrastructure and architecture aimed at improving flexibility and efficiency in a rapidly changing global economy. This is being enabled by ubiquitous network connectivity and rapid improvements in the price and performance of computing and network resources. There are significant advantages to be gained by organisations adopting some of the technologies now becoming available, and the prospect is of ICT being an enabler of new ways of working rather than an inhibitor of change. The real significance of the trends we have identified will become apparent as the service-based approach to infrastructure becomes widely adopted and allows the emergence of a global ecosystem of services, service providers and users, interacting via a pervasive ICT infrastructure.

References

DiDio, L. and Hamilton, G., 2006, Virtualization, Part 1: Technology Goes Mainstream, Nets Corporations Big TCO Gains and Fast ROI, Yankee Group.

e-Business Watch, 2007, The European e-Business Report 2006/07, http://www.ebusiness-watch.org/key_reports/documents/EBR06.pdf European Commission.

Heffner, R., 2006, Your Strategy For Web Services Specifications, Forrester Research.

International Standard, 10746–2, 1996, Information Technology- Open Distributed Processing- Reference Model: Foundations, ISO/IEC.

Jorgensen, D. W. and Wessner, C. (eds), 2007, *Enhancing Productivity Growth in the Information Age: Measuring and Sustaining the New Economy*, National Academies Press, Washington, DC.

OASIS (Organization for the Advancement of Structural Information Standards), 2006, Reference Model for Service Oriented Architecture 1.0, OASIS Open.

PCAST, 2007, Leadership Under Challenge Information R&D in a Competitive World.

Phelps, J. R. and Dawson, P., 2007, Demystifying Server Virtualization Taxonomy and Terminology, Gartner Research.

7

Achieving Security in Enterprise IT

Theo Dimitrakos and Ivan Djordjevic

7.1 Introduction

Cearley *et al.* (2005) made the following predictions about the use and integration of Service-Oriented Architecture and Web services across enterprises: 'Web Services Security and XML Key Management Specification will provide a relatively complete standards foundation for security; however, trust issues and the cost and complexity of the supporting infrastructure will constrain most Web services deployments to early adopters and low-value transactions' (80% probability).

The same report predicted that the dominant Web services security concern will shift from a focus on the lack of standards and mechanisms for securing Web services deployments to monitoring and control issues. With a nearly complete set of basic security standards in place, Web services perimeter products will begin to mature. Larger vendors will enter the marketplace or acquire smaller startups. Indeed, as reported in Infonetics (2007), World Wide Web services security and content gateway sales 'grew 8% between Q1 and Q2 of 2006, reaching $270 million, and are forecast to grow 43% by the Q2 of 2007. Annual worldwide sales are expected to hit $2.3 billion in 2009'. North America accounts for 52% of all content security gateway revenue in 2006, Asia Pacific for 25%, EMEA for 21%, CALA for 2%[1]. Furthermore, by the end of 2006, the SOA XML appliances market has been particularly dynamic with key vendor mergers and acquisitions being reported almost in every fiscal quarter.

---

[1] EMEA refers to Europe, Middle East, Africa. CALA refers to Central America, Latin America.

ICT Futures: Delivering Pervasive, Real-time and Secure Services
Edited by Paul Warren, John Davies and David Brown
© 2008 John Wiley & Sons, Ltd

So what is the big driver for this market? Most vendor strategists and market analysts would agree that the big driver is simplifying the infrastructure, integrating functions into a network-centric hardware appliance that is easy to operate.

The complexity and risk inherent in integrating systems in an SOA environment is overwhelming. SOA appliances offer an answer to questions customers contemplating SOA have been asking of their vendors for sometime: How do I deal with security? How do I deal with the complexity? How do I know that my service levels are going to be correct?

Most – if not all – of such SOA appliances are currently focusing on Web Services exposure and Layer 6/Layer 7 security (WS-* Security, XML security, Secure Socket Layer (SSL), Hypertext Transfer Protocol over SSL (HTTPS)), content inspection/ transformation, and content-based message routing. As their use of corporate SOA infrastructures matures and competition between vendors intensifies, we anticipate that a new, larger market will evolve that will be focusing on incorporating value-added services on such SOA devices in order to cater for commonly useful Business-to-Business functions, such as federated identity management, distributed access control, enforcement of service-level guarantees, enterprise application integration, real-time business intelligence (e.g. event-driven reconfiguration). Also, as deployments increase in complexity, we predict a stronger need for virtualising, clustering, and managing the SOA devices over a wide area network.

The Service-Oriented Infrastructure (SOI) Secure Service Gateway concept (SOI-SSG) developed at BT's Security Research Centre is addressing this emerging market. SOI-SSG builds on the integration of clusters of virtualised SOAed appliances with value-added services for trust federation, federated identity, access management, secure message processing, real-time adaptation, service-level guarantee enforcement, monitoring, and accounting. SOI-SSG is used throughout this chapter as an example of how security services will be provided in a service-oriented world.

In most common deployments, SOI-SGG takes the form of a managed (and potentially) network-hosted service that enables an enterprise in a network economy to achieve the following benefits:

- to virtualise its applications, employee accounts, and computing and information resources;
- to maintain the management of such virtualised entities, including defining trust relationships, security and access policies, identity schemes, etc. that apply for them in the scope of each B2B collaboration that they participate in;
- to adapt the observable behaviour of virtualised entities and the use of infrastructure services in response to contextual changes. Examples include updating the security, privilege provisioning, or access control policy in response to changes of business process activity, changes of membership in a B2B collaboration, changes of the location of the service requester;
- to securely expose such assets to an open network, with the option to apply different security policies in different collaboration contexts, while ensuring process and information separation between services transacting in different collaborations;
- to maintain the management of its participation in B2B collaborations. This includes managing the life cycle of its participation in a B2B collaboration.

7.2 Concept

7.2.1 Overview of SOI-SSG

SOI security solutions come as stand-alone value-added security services or are integrated into an SOI-based Secure Service Gateway (SOI-SSG) for securing the exposure and end-to-end integration of business applications within the enterprise and between business partners.

SOI-SSG offers a flexible, transparent, and manageable security layer to protect applications within and across the enterprise. Its extensible service-oriented architecture facilitates the use of other standards-based security capabilities offered from third party security service providers. SOI-SSG can also establish trust and can be used both for protecting the exposure of value-added services (such as BT's 21CN common capabilities[2] or some of the reusable services at Gridipedia[3]), including outsourced security services, and for integrating such capabilities into an interconnected SOI-SSG, therefore enhancing its security functionality. For example, BT can expose through an SOI-SSG trust establishment services, such as B2B federation management services, or a federated identity broker, and securely interconnect it to the SOI-SSG used by a BT customer, hence enabling the customers to use such trust establishment services as part of their security infrastructure. When used in conjunction with SOA-based service integration platforms, such as BT Integrate[4], SOI-SSG enables seamlessly integration of such value-added services into the Enterprise Bus.

Relative to current Web 2.0 and Web Services security products, the SOI-SSG enhances IT Operations and SOA Governance with improved policy management, and visibility, control of the granularity of security policy presentation, and context awareness. It offers connectivity to external identity and security attribute providers; it supports managing the full life cycle and federation of trust realms; it supports distributed access control; it offers real-time monitoring of security enforcement; it supports real-time security policy adaptation in response to events.

SOI-SSG can be virtualised in order to protect assets of different customers over the same infrastructure. SOI-SSG virtualisations can also be aggregated and managed as a gateway cluster in order to support high-volume transactions, and to assure service availability while under attack, at a single point or in a distributed enterprise.

Advanced SOI security capabilities that are integrated into the SOI-SSG, but can also be offered as network-hosted security services, are;

1. *Virtualisation* and *life-cycle management* services that manage the life cycle of a secure exposure of enterprise resources. Enterprise resources are virtualised by associating a service-oriented infrastructure configuration and a set of policies for the relevant infrastructure services with a service end point that is uniquely identifiable within a distinct collaboration context.
2. *B2B collaboration management* services that support the full life cycle of defining, establishing, amending, and dissolving collaborations that bring together a circle of

---

[2] See http://www.btplc.com/21CN/ for more information about BT's 21C Network.
[3] See http://www.gridipedia.eu/ for the Gridipedia web site.
[4] See http://www.bt.com/uk/bt_integrate/.

trust (federation) of business partners in order to execute some B2B choreography. The lifecycle of the B2B collaboration is divided in four phases: *identification, formation, operation*, and *dissolution*. See also Dimitrakos T., Kearney P., Goldby D. (2004).

3. *B2B Federation services* and *security token services (STS)* that allow managing the full life cycle of circles of trust, identities, and security attributes within and across enterprises. The functionality for generating security tokens (including the token format) is extensible, supporting custom XML tokens, SAML[5] assertions, and X.509[6] certificates at the same time.

4. *Distributed access control (Policy Decision Point, PDP)* and *authorisation services* that allow the necessary decision making for distributed enforcement of access policies by multiple administrators, ensuring regulatory compliance, accountability, and security audits.

5. A *secure message processing system* that allows protecting XML and Web services messages in B2B transactions, by enforcing content- and context-aware message processing policies on a virtualised resource.

6. *Security autonomics* implementing novel technology that allows reconfiguring the security services in response to security or Quality of Service events in order to optimise performance, to respond to threats, and to assure compliance with agreements and enterprise policies.

7.2.2 Life-cycle Model

The lifecycle of a secure B2B collaboration can be broken down into a number of stages, as presented in Figure 7.1. In the following subsections, we focus on common functions that can be offered through the SOI-SSG and underpin the B2B collaboration life cycle:

- Define Policies and Agreements. This stage covers the definition of federated identity schemes, security and access policies or policy templates, event structures, and collaboration agreements. Overall, this stage is about agreeing how to collaborate.
- Federate. This stage covers the choice of identity providers, the formation of a specific circle of trust, the specialisation of access policies and collaboration agreements, and the correlation of internal and external identities and attributes. Overall, this stage is about establishing a concrete collaboration context.
- Virtualise. This stage covers the secure exposure of a resource within a B2B or B2C collaboration context (viz. 'service virtualisation') and the specialisation of SOI capabilities (identity services, authorisation services, message processing engines, etc.) in order to enable the exposure (viz. 'security infrastructure virtualisation'). Overall, this stage is about configuring the SOI so as to support end-to-end service transactions and integration within the context of a federation.
- Enforce. This stage covers the operation of the SOI infrastructure that underpins the B2B collaboration. It is about intercepting, processing, transforming and routing

---

[5] Security Assertion Markup Language, an XML standard for exchanging authentication and authorisation data between security domains.
[6] X.509 is an ITU Telecommunication Standardization Sector (ITU-T) standard for public key infrastructure.

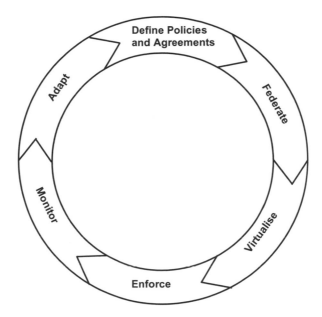

Figure 7.1   B2B collaboration lifecycle

messages, recognising collaboration contexts and enforcing the appropriate security and QoS policies, and the contextualised invocation of the appropriate SOI common capabilities. Overall, this stage is about ensuring end-to-end security and enabling the secure execution of service transactions within the B2B collaborations.

- Monitor. This stage covers the interception and monitoring of transactions between resources in a given context, constraining information sharing, and presenting relevant information in dashboards (e.g. policy successes/violations, utilisation of resources and SOI capabilities, state of transactions, etc.). Overall, this stage is about analysing the state of SOI-SSG instances, in particular, and of the B2B collaboration in general.
- Adapt. This stage is about the SOI security capabilities reacting to changes or implementing changes in order to meet business objectives. It covers the correlation of events that raise alarms or provide other means of input to real-time-business intelligence mechanisms and the enforcement of reactions that bring about a reconfiguration of the SOI security capabilities.

7.2.3 Define Policies and Agreements

The objective of this stage is to offer collaborative SOA governance via B2B agreements between cooperating units and partners. This is achieved by SOI establishing a collaborative governance plane that allows a group of administrators to collaboratively manage a view of the SOI that enables B2B transactions and ultimately the integration of resources exposed by each unit in a given collaboration context. Policy-based, context-[7] and content-aware[8] security management and access control combined with the constraint delegation of administrative authority offer flexibility and increased control over collateral and shared assets and improve the information flow management between collaborations and collaborators.

The SOI-SSG architecture ensures the contextualisation of security policy and security capabilities,[9] and enables multiple administrators in an enterprise to efficiently manage security capabilities over a common SOI. It also allows for different choices of identity provider, authorisation service, monitoring policy, exposure gateway, and administrative delegation scheme in different collaboration contexts. Furthermore, such choices are transparent to the rest of the collaborators, as long as they comply with the B2B agreements and build on a common set of standards. Hence, an enterprise can expose behaviourally different services that are built by composing the same business function (in terms of its core application logic) with behaviourally different value-added services and security capabilities through the SOI-SSG.

7.2.4 Federate

The objectives of this stage is, on the one hand, to create a circle of trust that enables identity federation and, on the other hand, to configure the SOI-SSG of a partner in order to enable execution of the B2B transactions and to enforce security and QoS in accordance with the enterprise policy of each partner and in compliance with a concrete collaboration context.

Through the use of Web 2.0 and Web services technologies, enterprises seek standardised mechanisms for communicating, enacting business processes, and sharing information across organisational boundaries. To maintain accountability and to ensure that only authorised users gain access to restricted operations or information, the exposed services must be able to communicate, identifying information across these organisational boundaries. SOI-SSG allows a loosely coupled integration of one's security and access services with multiple sources of identities and security attributes within a common federation context. It can support multiple sources of federated identities and security attributes, for resources as well as attributes and identities of

---

[7] Examples of context-aware policies include (i) the fusion of role-based and process-based access management in order to achieve specialisation of access control policies for a type of process (e.g. purchase order, collaborative engineering design, etc), (ii) the specialisation of identity provisioning (or access control) policies to the type for a collaboration (e.g. type of supply chain, virtual organisation, collaborative project, etc.).

[8] Examples of content-aware policies include policies for processing message security, whose enforcement is driven by the contents of the intercepted messages. That is, different policy assertions are enforced depending on whether the payload of an intercepted message complies with WS-Security standard, whether it uses a SAML token or a X509 token, whether it complies with the Web Service Definition Language (WSDL) of the requested service, the arguments of a service operation are within certain limits, etc.

[9] For example: Access policies can be parameterised by context descriptors. Federated identity providers (e.g. STS) can issue tokens containing different security claims depending on the identity federation context within the scope of which the token is provisioned and has to be validated. Events can be correlated based on context, etc. Also, service end points can be referenced in relation to a context (address scoping) and can be associated with different messaging policies and usage agreements for different contexts.

the services encapsulating business functions. The SOI-SSG architecture enables a clear distinction between internal and federated identities, on the one hand, and the source of attribute or identity provisioning, on the other hand. Notably, the STS summarised in Section 7.2.1 allows a choice of internal-to-federated identity bridging, and choices of internal and federated identity token formats, depending on the federation context for which the token is issued. Furthermore, extensible security tokens include security assertions that can be provided from different sources including, for example, business bureaux, government agencies, and mission-specific attribute authorities.

Some of the notable offerings are the following:

- Each partner in a given collaboration has a single logical source of federated identities; attributes from multiple sources can be composed into standardised security tokens; internal and federated identity are bridged through extensible context-sensitive schemes.
- Supports a selection of schemes for bridging internal and federated identities, and by facilitating control of the structure and type of the information conveyed in security tokens and security claims, SOI-SSG allows security managers to choose the degree of anonymity, the mixture of identity-/role- or attribute-based access, and consequently, the level of nondisputable accountability of actions (e.g. end-to-end nonrepudiation) that they want to enforce. In turn, this helps in achieving the right balance between security and privacy for each business scenario.
- Supports policy compliance for segregation of duties and process isolation by contextualising identities and access policies.
- Controlled outsourcing of identity provisioning and brokerage, and choice of identity providers for each collaboration context, by virtualising and segregating SOI security configurations for distinct federation contexts.

The SOI-SSG architecture empowers each unit to manage the attributes and identities of their own resources, reducing the overhead associated with managing accounts for temporary personnel or the customer base of a business partner. It also facilitates coordinating reconfigurations of a circle of trust such as accepting new units/partners or removing existing units/partners by following a schedule (e.g. distinct tasks in a collaborative process) or with immediate effect.

7.2.5 Virtualise

This stage covers two conceptually analogous fundamental aspects of the SOI-SSG:

- *Security infrastructure virtualisation.* SOI-SSG enables a partner to virtualise – at real time – the SOI security infrastructure (including any network-hosted security capabilities offered through the SOI-SSG), and to associate such infrastructure virtualisation with a context, such as the context of a concrete collaboration, as explained in Section 7.2.4 [such a context can be denoted by a WS-Federation (OASIS) context identifier]. All application virtualisations taking place in the scope of this context will rely on a common infrastructure virtualisation.

- *Application virtualisation.* To securely expose – at real time – a resource in a given collaboration context and to subsequently manage the specialisation of the SOI-SSG configuration including all associated security policies for the life cycle of such exposure.

At any time during the life cycle of the exposure, the security autonomics engine or another authorised client may request the reconfiguration of an SOI capability involved in this exposure. Reconfiguration entails updating the policy associated with the SOI capability and, once the update is successful, synchronising the internal policy repository and confirming the successful reconfiguration, otherwise raising an alarm.

Upon an authorised request, the SOI security infrastructure may terminate an exposure. Terminating an exposure includes archiving the SOI security configuration associated with this exposure (including the addresses of the relevant SOI security capabilities and their resource and federation context-specific policies) and updating/destroying the state of the corresponding SOI security capabilities relating to the exposure, and releasing resources that have been reserved.

The same resource can be virtualised multiple times in different (or even the same) federation contexts. The life cycle of distinct exposures can be managed simultaneously by the SOI security subsystem.

7.2.6 Enforce

This stage covers another fundamental aspect of the SOI-SSG operation. It enables the infrastructure to intercept, process, augment, and protect message exchanges between exposed business functions based on their content and the federation and transaction contexts. It also enables the infrastructure to invoke value-added services such as network-hosted SOI security capabilities offered through the SOI-SSG in order to enforce security policy decisions, to validate, to transform, and to route messages depending on their content and the federation or transaction context.

Both incoming and outgoing messages are intercepted. Security policy enforcement and invocation of value-added services always depends on the content of the intercepted message and the context of the interception – e.g. the semantics of the message exchange pattern, the security state, and the scope of the transaction or federation within which the message exchange pattern is implemented. Invocation of value-added services takes place during the processing of intercepted messages, and the validity of the policies including evaluation against the administrative delegation constraints in place is established at real time.

7.2.7 Monitor

This stage covers the capabilities enabling the infrastructure to inspect, analyse, and visualise information (to authorised operators and security administrators) relating security policy, the state of its enforcement, and its relation to the business objectives.

The infrastructure inspects the meta-data of the messages exchanged between business services as well as the messages exchanged between the SOI security services, extracts, and potentially anonymises information that is then included in notifications to services specialised in monitoring and analytics. Monitoring in a large-scale infrastructure requires a topic-based, publish-subscribe, notification mechanism that is able to reliably and securely disseminate information in accordance to monitoring and audit policies.

The SOI-SSG architecture focuses on the following aspects of monitoring:

- *B2B security dashboard*: real-time monitoring of security state throughout the value-chain, based on privacy/confidentiality preserving security information sharing between a partner and its collaborators.
- *Granular security monitoring and compliance assurance*: real-time security monitoring to identify violations of specific clauses of a security policy or a B2B agreement for a given service (or group of services) in a given transaction (or a business process) in a given federation context.
- *Contextualised visualisation of security service interactions*: visualisation of the interactions within an SOI-SSG domain. An SOI-SSG domain may span across a service provider using the SOI-SSG and the providers of the value-added services offered through the SOI-SSG. The visualisation shows the history of each transaction, and allows the multi-modal analysis[10] and visual exploration of important meta-data.

7.2.8 Adapt

This stage covers capabilities enabling the SOI-SSG to automatically[11] reconfigure itself in reaction to a system event or a high-level administrative decision.

Adaptation provides a basis for a coordinated reconfiguration of the SOI-SSG in response to state of business transactions, security alerts, performance, resource availability, and business impact of security operations.

Adaptation in response to events is driven by contextualised event-condition-action policies (ECA) that (i) correlate and evaluate SOI events in accordance to policy constraints and (ii) trigger SOI security actions depending on the result of the evaluation.

For adaptation to work at real time, the SOI security capabilities offered through SOI-SSG expose programmatically manageable interfaces, and SOI-SSG comes with adaptive administrative processes to coordinate updates to the contextualised security state that is distributed among several SOI security capabilities.

---

[10] 'Multi-modal analysis' means enabling the application of (a) multiple content recognition and analysis mechanisms over a common set of events, (b) multiple context-based correlation algorithms over a common set of events, (c) spatial (e.g. location-specific) and temporal patterns based for analysing events against usage-control policies. 'Multi-modal visualisation' means enabling the utilisation of multiple complementary forms of presentation of the results of each analysis.

[11] Obviously automatic reconfiguration in reaction to events covers a subset of what can be manually reconfigured at real time through the intervention of an authorised administrator.

7.3 Scenarios and Case Studies

7.3.1 Supply Chain: VO Management in Collaborative Engineering

This case study shows a number of small and medium enterprises in the engineering and scientific sectors that federate to form a secure collaboration [a collaborative engineering virtual organisation (CE VO)]. The collaboration permits new members/ suppliers to enter or leave the collaboration without any disruption to the operation of the CE VO. This case study has been extensively analysed by a team from British Aerospace Systems in the scope of the TrustCoM collaboration. See also Dimitrakos T., Kearney P., Goldby D. (2004) and Dimitrakos T., Ristol S., Wilson M. (2004).

In the scenario (see dark arrows in Figure 7.2, the analyst needs to retrieve (1) the design from the product database, stores (2) this on the storage provider which returns an identifier for this bulk of data. The analyst then submits this identifier together with an activation request (3) to a service managing the high-performance computer platform. Now this service retrieves (4) the input data from the storage provider, computes the calculation, and stores (5) the output data on the storage provider under the same identifier. Using the identifier, the analyst can retrieve the output of the simulation for further processing.

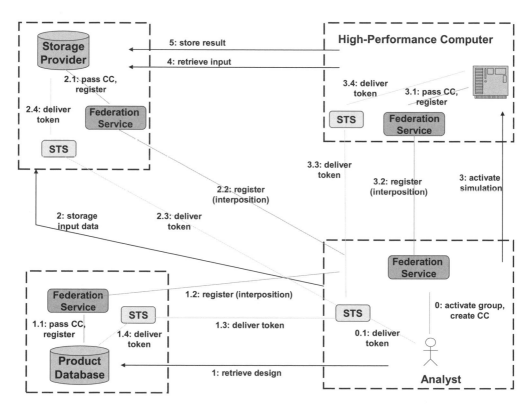

Figure 7.2   Secure group used for collaborative engineering

Through the use of SOI-SSG, participants can automate initiation and management of a secure collaboration, as depicted in Figure 7.2 and as described below. A demonstration prototype has been implemented in the context of the TrustCoM project[12].

7.3.2 Entertainment Services: Online Gaming/Virtual Music Store

Here we present a case study of the B2B gateway applied to the design and development of a virtual hosting environment (VHE) for online gaming. This is a sector-focused pilot project being carried out in the context of the BEinGRID programme[13]. However, we envisage that the VHE paradigm will be suitable for use across a number of sectors and a variety of business scenarios requiring remote hosting or service outsourcing over a flexible ICT infrastructure.

Virtualisation of hosting environments refers to the federation of a set of distributed hosting environments for execution of an application and the possibility to provide a single virtual access point (e.g. the B2B gateway) to this set of federated hosting environments. In a typical scenario, a number of host providers offer hosting resources to the application providers for deploying and running their applications, which are then 'virtualised' with the use of middleware services for managing nonfunctional aspects of the application, and are transparently exposed to the end user via a single VHE.

Currently, the gaming platform is static in nature, with dedicated game servers, resulting in extreme peaks in demand, due to the period of the day or week. This causes very low average utilisation of dedicated gaming servers, and therefore high cost of initial investment and maintenance. One of the main challenges that the gaming provider is facing is to make available on-demand high-performance servers for game execution and to utilise advanced statistics and user community management capabilities, offered by different providers.

The approach taken through the use of the VHE is to make available infrastructure services for security, community management, and virtualisation that can be used by various service providers, allowing them to link with each other. This is achieved via a B2B gateway offered by an infrastructure provider. By making available these infrastructure services for the use of various stakeholders, they are enabled to federate and engage in various advanced models of cross-enterprise business collaborations. Consider the case depicted in Figure 7.3, which is being addressed in the pilot, as a realistic example of a business model enabled with the VHE.

In this scenario, the game application provider deploys its gaming application onto two different execution environments (gaming servers), owned by different host providers. The game platform provider, who wants to offer the game to an end user, discovers gaming servers and creates business relationships with them, and also with a separate service provider who offers a system for community management (of gaming clans, tournaments, advanced statistics). Through the use of the VHE, these various services are offered transparently to an end user, including the game platform provider's ability to perform the load balancing and server selection based on the defined SLAs.

---

[12] http://www.eu-trustcom.com/.
[13] http://www.beingrid.eu/.

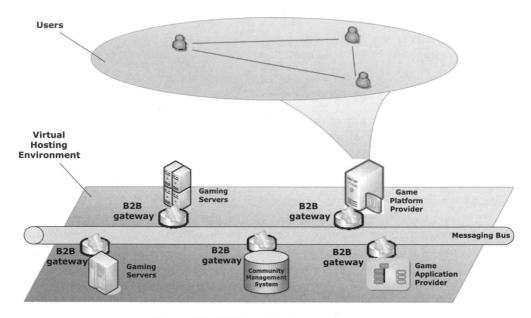

Figure 7.3   VHE-enabled gaming scenario

7.3.3 Defence/Crisis Management

This scenario covers the setting up of a (communication) configuration to support a
military task force. Its principal purpose is to illustrate how the security architecture
can be used to deploy infrastructures that are both secure and flexible. Such infrastructures include services that are provided by different organisations (including civilian
contractors). The services accessed can be raw communications channels, but can also
be suppliers of data such as satellite reconnaissance, intelligence data, technical databases, etc. Each one will need to be accessed on a need-to-know basis.

Upon the identification of a need to deploy forces, the appropriate position in the
chain of command (dependent upon size of deployment) forms a network-enabled
coalition. This involves:

- selecting various units and third-party contractors who will constitute a virtual
  organisation and
- selecting policies that control the behaviours of the virtual organisation.

These policies include identity provisioning, information disclosure, access policies
to services in different trust realms, and can be applied to both incoming and outgoing
interactions within a federation of trust realms. The collection of policies to be applied
may include policies covering multiple layers of the network stack, e.g. from setting
up channels for priority traffic to protecting messages, to protecting access to services,
to protecting access to information that has been exchanged. Policies may evolve
within the lifetime of the virtual organisation: they can be enriched, updated, or
revoked.

The selected units do not have to know the identity of the other units/organisations taking part in the virtual organisation – e.g. civilian contractors do not need to know about the size of a deployment and a need-to-know basis can be maintained for disclosing identity attributes and privileges of the units operating in-theatre.

When a target of opportunity is identified, the configuration must be rapidly adjusted. This involves swiftly establishing circles of trust between particular HQs and the relevant units. Administrative rights may need to be temporarily delegated for certain actions to specific HQ or personnel, which is taken care of by a constraint delegation framework of the PDP (it allows secure management of such delegation actions; it establishes their authenticity and integrity; it authenticates their issuer; it constrains their duration and purpose, and allows to historically audit delegation actions ensuring that administrators are accountable for their actions, even if a level of anonymity is assured for delegation actions outside of their trust realm).

During the deployment, the networks can come under a denial of service attack. This can be detected by a security autonomics sub-system that, in response to events, can trigger an adaptation of the policy model and security architecture in place through the ECA policies (set up to respond to events or patterns of events). Detection of an attack may result in reconfiguration, e.g. a policy change, change of virtual address of a resource, change of load sharing in a gateway cluster, change of or denial of access to attackers.

Maintaining availability under attack and survivability of mission-critical services and data is critical in defence coalition scenarios, especially in order to maintain an appropriate quality of service for legitimate units and to localise the impact of the attack as the attack unfolds and until the attackers are identified and dealt with.

With a long deployment, it may be necessary to switch units in and out of the virtual unit, or to introduce extra networked resources, e.g. information stores. Finally, when the virtual unit is disbanded, the accounting service needs to retain information on actions taken for later audit.

7.4 Conclusion

In this chapter, we have provided an overview of security concepts, models, and technologies for the service-oriented enterprise. We used an innovative SOI security framework developed by BT Research in order to (a) illustrate, on the one hand, how these concepts and technologies can be combined together in order to achieve security in IT-driven business environments and (b) to offer an example of how security services will be provided in a service-oriented world. Our intention has been to explore service-oriented architecture as a foundation for offering a flexible and manageable security infrastructure and to propose the result as a means of securing the service-oriented infrastructure that will underpin business collaborations between the service-oriented enterprises of the future.

References

Cearley D.W. *et al.* 2005, *Gartner's Positions on the Five Hottest IT Topics and Trends in 2005*. Gartner Research Report, May 2005. ID Number: G00125868.

Dimitrakos T., Kearney P., Goldby D. 2004, *Towards a Trust and Contract Management Framework for Dynamic Virtual Organisations*. In P. Cunningham and M. Cunningham (ed.) eAdoption and the Knowledge Economy, 2004 IOS Press, Amsterdam.

Dimitrakos T., Ristol S., Wilson M. 2004, *TrustCoM – A Trust and Contract Management Framework enabling Secure Collaborations in Dynamic Virtual Organisations*. ERCIM News No. 59.

Infonetics 2007, *Content security gateway appliance revenue jumps 70% in 2006*. Infonetics Market Research report, March 2007.

8

The Future All Optical Network – Why We Need It and How We Get There

David Payne

8.1 Introduction

Future content rich services will continue to drive bandwidth growth in telecommunications networks, and increasing demand for these services will be a major driver of fibre into the access network, if ways to economical deployment can be found. If fibre does penetrate access networks on a large scale, then there will be huge bandwidth growth throughout the network hierarchy: access, metro and core networks.

Future growth driven particularly by increases in video content with higher definition standards, eventually moving beyond today's highest high-definition television (HDTV) resolutions, coupled with user behavioural changes such as increasing personalisation of programming material, distributed storage and distributed content systems, such as peer-to-peer file sharing, could see average user bandwidths going beyond 10 Mb/s and peak rates requiring several hundred Mb/s to individual users.

These levels of bandwidth demand could drive two or three orders of magnitude growth in network bandwidth over the next 10 years. Digital subscriber line (DSL) and cable modem technologies with today's hierarchical network architectures will not be able to cope or scale economically to meet these unprecedented demands.

In this chapter, the role of all optical networking to meet these future demands will be briefly reviewed, why we will need it, the drivers and the architectural changes that will be required to enable future networks to remain economically viable and to be able to scale to the enormous bandwidths that could arise in the future.

ICT Futures: Delivering Pervasive, Real-time and Secure Services
Edited by Paul Warren, John Davies and David Brown
© 2008 John Wiley & Sons, Ltd

8.2 Why Optical Networks?

8.2.1 Physics Considerations

The history of the development of communications technology has seen an increasing exploitation of the electromagnetic spectrum whereby higher 'carrier' frequencies have been used to enable ever greater bandwidth modulation of information on to them. The current generation of optical communications technology uses optical fibre operating in the ~1600 nm to 1260 nm wavelength range, a 'carrier' frequency range from 187 THz to 238 THz.

It might be supposed that as technology progresses, we could exploit higher and higher 'carrier' frequencies indefinitely. However, the photonic nature of electromagnetic radiation means that the photon energy is also increasing in proportion to the carrier frequency. The problem with higher photon energies arises from the quantised nature of the universe and the uncertainty in the number of photons arriving at a detector in a given interval of time. This uncertainty produces a fundamental noise in communications systems called quantum noise and becomes the limiting factor to the information carrying capacity of a channel as higher frequencies and photon energies are exploited.

At low frequencies, thermal noise dominates and is the fundamental limit to channel capacity. As frequencies approach optical frequencies, quantum noise becomes more significant and at ultraviolet and beyond, quantum noise begins to severely limit channel capacity. The effect of quantum noise on channel capacity is shown in Figure 8.1.

The channel capacity curve assumes that a constant percentage of the carrier frequency can be used for the information bandwidth that is modulated onto the carrier. Using this assumption it can be seen that the channel capacity peaks around 5×10^{16} Hz (~3 nm wavelength). The roll-off in information capacity beyond 3 nm wavelength is due to the dominance of quantum noise and is sufficiently great to produce an effective ceiling on information capacity as illustrated by the cumulative capacity curve.

What is also interesting to observe is that today's optical fibre technology is only an order of magnitude or two from this ceiling so, we can't expect to see the same enormous capacity gains in the future as we have seen in the past when moving from radio frequency communications to optical frequency communications. In fact, although it is possible that new, even higher speed technologies that operate between today's near infrared wavelengths and the far ultraviolet/near X-ray region may emerge in the future, there is currently no sign of these technologies being researched for communications purposes anywhere in the world.

This means that as fibre penetrates to all parts of the network (access, metro and core or backbone networks), the same intrinsic technology is being used and the same fundamental capacity is available in the access network as in the core network. This is the first time in the history of communications network that this has happened. In the past, there has always been a higher capacity transmission technology for use in the core network that traffic from the access transmission media could be multiplexed into.

Multiplexing of traffic from the access network into core transmission systems will have to continue with optical networking if communications networks are to remain

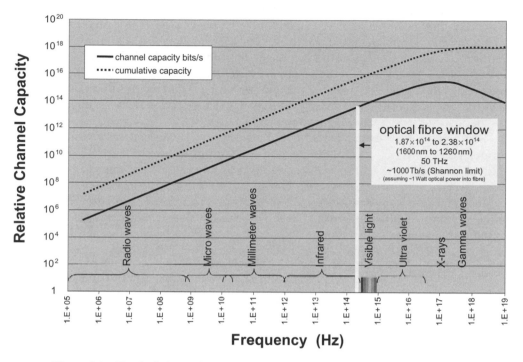

Figure 8.1 Physical channel capacity limits as a function of 'carrier' frequency

practicable and economically viable. Therefore, if the core technology is optical fibre and the access technology is also optical fibre, it is implicit that the end users connected to the access network via fibre must be sharing the total capacity of the fibre with other users. Individual end users cannot have all the capacity of the optical fibre dedicated to them because there is no higher capacity transmission technology to multiplex fibre capacity into.

8.2.2 Commercial and Service Considerations

There are a number of drivers for operators to install optical fibre in the access network. These include:

- meeting competitive threats;
- reducing operational cost;
- meeting end user demand for new high bandwidth services;
- staying internationally or regionally competitive;
- new revenue generation.

Once a fibre to the premises (FTTP) network is installed, there is no competing network technology that can outperform it in terms of technical capability. There may be performance differences, due to the choice of equipment and architecture, in terms of service quality and economic viability, but the physical fibre infrastructure cannot be bettered by any currently known technology. This begs the question whether competition at the physical infrastructure layer has any real meaning or advantage in a fibre future, especially when the economic payback is difficult enough with only one network to finance; more than one may not be viable. This suggests that there may be a shift from competition at the physical layer to more competition at higher layers, maybe just the service and application layers.

There is now good evidence that fibre networks are intrinsically more reliable than copper pair networks, leading to much lower fault rates and reduced network visits leading to fewer intervention induced faults. Also, because fibre can support all services, service provision and service churn can be automated processes only rarely requiring an end user or network visit. With properly engineered end user and service management systems, the vast majority of service changes and provisions should be configurable remotely, including churn from one service provider to another. It has been argued that operational savings alone could justify FTTPs being installed (Halpern *et al.*, 2004).

As new high bandwidth requirements emerge, the demand for faster access speeds from end users and their impatience and dissatisfaction with slow networks could come to the fore. The problem of slow access speeds is further illustrated in Figure 8.2, which shows the time taken for large files to be downloaded via various technologies and access speeds. For large files such as video files or collections of high resolution images, etc., only FTTP comes close to delivering short enough transfer delays that won't severely test the user's impatience. Fibre to the cabinet is not adequate.

This demand for much faster access speeds (not necessarily huge increases in sustained bandwidth) is putting pressure on operators and administrations to put high

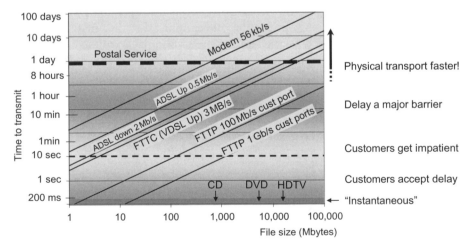

Figure 8.2 Time to transmit files. FTTC, fibre to the cabinet (ADSL, Asymmetric Digital Subscriber Line; VDSL, Very High Speed Digital Subscriber Line)

bandwidth FTTP networks in place. This may be especially so if there are neighbouring competing regions offering FTTP, and by doing so attract inward investment, boosting local economies, employment, house prices, etc. This certainly appears to be one of the drivers for municipalities who are willing to invest in basic fibre infrastructure for the benefit of their local communities, and indeed there does seem to be some evidence that FTTP can provide these local advantages (Lehr *et al.*, 2005). The concept of staying internationally competitive through deployment of the best telecommunications infrastructure is a major driver for Japan and Korea, and there is growing concern in the US that America is falling behind and is not the technology leader it once was (Prestowitz, 2006). The slow adoption of FTTP in Europe may well mean it will fall behind both the Far Eastern economies and the US if it fails to grasp the opportunities in the next few years.

The last driver listed above is new revenue generation. This is an area of major uncertainty; although FTTP will certainly enable new high capacity services, it is unclear whether there will be any significant revenue growth over and above traditional trends. The problem is that revenue generation derived directly from new IT and bit transport services are often substitutional, i.e. new service revenue, simply displaces legacy service revenue and net revenue growth remains relatively static.

Historical analyses of revenue growth and bandwidth price decline suggest that there may not be sufficient revenue growth to sustain traditional network builds and that radically new architectures will be necessary to change the end-to-end cost structure of networks and massively reduce the cost of bandwidth provision (Payne and Davey, 2002).

8.2.3 Bandwidth Growth Scenario

The major problem arising from an FTTP future will be the huge level of bandwidth growth that this technology can support and deliver into the metro and core networks. By using service scenarios and traffic scenario models, it is possible to get a feel for the level of bandwidth growth that may arise in the future (Payne and Davey, 2002). It should be stressed that these models are scenarios to aid network strategy decisions and are not forecasts for network plan and build purposes.

Results from service scenario models show that average bandwidth per user could approach 10 Mb/s per user, and the access connection speed will need to be a minimum of 100 Mb/s and ideally greater than this to support the much higher instantaneous rates required.

The only practical technology to be able to offer such bandwidths and connection speeds is FTTPs coupled with optical networking in the core. This will be discussed further in later sections.

8.2.4 Burst-mode Operation

The simple models that are often used to generate bandwidth usage forecasts generally assume that video services will be streamed to the end user at the viewing rate. While this is true for live broadcast material, an alternative method of transmission can be

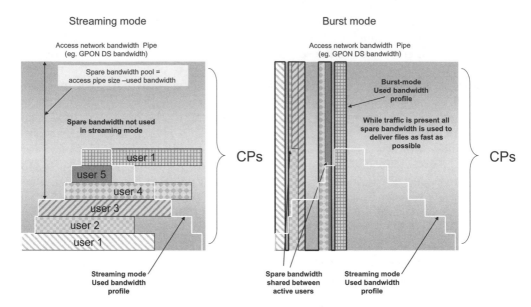

Figure 8.3 Comparison of burst-mode operation v streamed mode (CP, Communications Provider; DS, Downstream)

used for stored (pre-recorded) video files. That is to treat all pre-recorded video files as data objects and to burst the delivery to end users as fast as possible using any spare available bandwidth. This 'burst-mode' transmission approach would enhance the end user experience and wouldalso benefit network operation by removing long duration traffic sessions from the network as soon as possible, freeing up network resource for new traffic arriving.

The principles of burst-mode operation are simplistically illustrated in Figure 8.3. It shows end user video sessions arriving at a shared access connection and being handled by the network as conventional streamed traffic and alternatively as burst traffic.

In conventional streamed mode, bandwidth is allocated to end users as their traffic is admitted to the network. The level of bandwidth allocated is determined by the streaming rate required and agreed for that service. Any spare bandwidth on the access pipe is unused. As sessions are added, more bandwidth gets allocated and the unused bandwidth changes in line with sessions arriving and leaving the network. But to ensure an acceptable grade of service for the users, there is always some unused bandwidth available.

In burst-mode operation, a session is accepted under the same minimum bandwidth requirements as in the streamed mode case. However, if there is any unused bandwidth available, this is allocated to the active users using some bandwidth allocation policy. For the example shown in Figure 8.3, an equal share over active sessions is used. When the first session arrives, the streamed case allocates the streaming bandwidth allocation. However, in the burst-mode case, there is only one user active and all the available bandwidth is allocated to this first user. By the time the second user traffic arrives, in this example, under burst-mode operation, the first user session has completed and therefore this second user also gets all the pipe bandwidth allocated. When the third user session starts, the second user session has not completed, and the access bandwidth is allocated equally between the two active users. When the second user session completes, all the bandwidth get re-allocated to the remaining third user whose session completes well before the fourth user session starts.

What is happening under burst-mode is that whenever possible, the maximum available bandwidth is always used to transmit any files on the network. This removes traffic from the network as fast as possible and maximises the bandwidth available to handle any incoming traffic demands. Also, the end user is experiencing much faster delivery speeds and a much more responsive system. If the end users have fast user ports with very high-speed access connections, then very high burst speeds of several hundred Mb/s can be achieved.

8.3 FTTP Technology Options

Operators and administrations wanting to deploy FTTP to the mass market have two general network architectures to choose from: point-to-point networks or shared networks using passive optical network (PON) principles. The main contending architectures for FTTP today are shown in Figure 8.4.

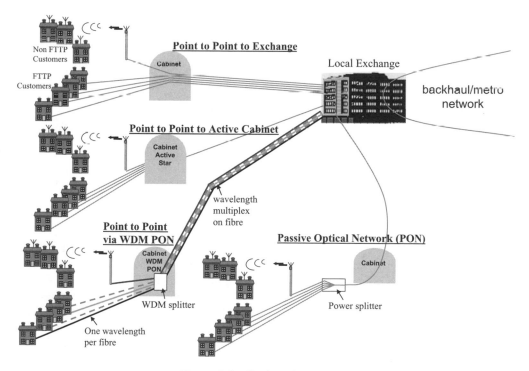

Figure 8.4   Options for FTTPs

8.3.1 Point-to-Point Fibre Solutions

The 'obvious' architecture for FTTP is a direct analogue of the ubiquitous copper pair telephony network and has a dedicated fibre (or fibre pair) from the local exchange/central office to the end user (see Figure 8.4: Pt to Pt to Exchange). The advantages of this architecture are the following: it is simple, relatively secure, uses simple optical transceivers; in principle, each path can be independently upgraded, and each end user appears to have access to all the fibre bandwidths.

In practice, these advantages are not all realised. The point-to-point nature leads to high fibre count cable; several thousand fibres per cable will be required as the cables converge onto the local exchange. These large cables lead to greater duct congestion and can significantly increase installation and operational cost. Upgrades to one end user may affect other end users connected to the same common interface cards causing disruption to customers who do not want the upgrade.

The access connection is dedicated to a single end user, and bandwidth cannot be shared or flexibly assigned to other users. The huge fibre bandwidth cannot be allocated to an individual end user because of the reasons outlined in the Physics Considerations section. The traffic from many access fibre connections must be multiplexed into a fibre of the same basic type in the back haul and core networks for reasons of economic viability and practical engineering. Therefore, in practical systems, individual end users can only be offered a fraction or share of the fibre bandwidth.

To avoid the high fibre count cables entering the exchange/central office, an alternative option is point-to-point fibre to a street node closer to the end users (see Figure 8.4: Pt to Pt to active cabinet). In the UK network, this node would conveniently be the cabinet or 'primary cross-connect point', it typically serves about 300 end users and is usually less than 1 km from their premises. At the cabinet, multiplexing equipment aggregates traffic onto a feeder fibre (or fibre pair) which backhauls the traffic to the local exchange site. Because the feeder fibre count is much smaller, the cable sizes to the local exchange are correspondingly smaller. It may also be possible to economically take them deeper into the network before further multiplexing or switching is required which could enable bypass and eventual closure of the local exchange site, thereby reducing costs. The major disadvantage of this solution is the large number of powered and equipped street cabinets required, which would push the number of active electronic nodes from about 5600 local exchanges to about 80,000 active cabinets. This is certain to increase operational cost, but of course, it is no worse than some of the proposals for active street cabinets to house VDSL equipment or street multi-service access nodes.

Another approach for a point-to-point architecture that avoids large fibre count cables is the use of wavelength division multiplexing (WDM) technology in the access network. Although the base concepts were around in the mid 1980s, access WDM has received renewed interest in recent years. The objective is to try to exploit the passive nature and lean fibre use of the PON architecture by sharing the fibre infrastructure using a wavelength from the exchange location to each network termination. It is logically a point-to-point solution, dedicating a wavelength channel between the exchange building and the end user termination (see Figure 8.4: Pt to Pt via WDM PON).

8.3.2 PONs

PONs were developed to tackle the problem of high fibre count and the large number of opto-electronic devices required for point-to-point solutions. They do this by using passive optical splitters as optical multiplexing devices mounted in fibre splice housings in footway boxes or manholes. The passive splitters enable the capacity of the feeder fibre from the exchange to be shared over a number of end users; it also shares the exchange termination and the associated opto-electronic devices and electronics (see Figure 8.4: PON).

The advantages of the PON solution are smaller cables sizes and reduced number of opto-electronic components, all leading to lower cost. The protocol that directs and marshals the end user traffic and avoids collisions also assigns the PON capacity across the end users connected, using a process called 'dynamic bandwidth assignment' and enables high utilisation of the PON capacity. The optical splitters also enables broadcast services to be provided very simply, the appropriate service being selected at the end user optical network termination (ONT).

There are three PON systems available: gigabit passive optical network (GPON) is the Full Service Access Network (FSAN) PON Standard from the International Telecommunications Union (ITU) and supersedes the earlier Broadband PON (BPON) standard. There is also Ethernet passive optical network (EPON) (also called gigabit EPON or GEPON) which is an ethernet only version from the IEEE standards body. The GPON specifications cover a wide range of bandwidth possibilities but the version of interest to most operators is 2.4 Gb/s downstream and 1.2 Gb/s upstream. EPON or GEPON is the most widely deployed solution in Japan and Korea; the current standard is a 1 Gb/s system a gigabit ethernet payload, although 10 Gb/s systems are being developed; see Payne *et al.* (2006) for a fuller description.

All the above systems have potential for exploitation in the FTTP arena; however, they also all have some major economic challenges for mass market deployment. Part of that challenge arises because they only impact the access network architecture and do not address the problem of unprecedented bandwidth growth, arising from the FTTP networks and needing transport through the backhaul and core networks.

8.4 Long-reach Access

8.4.1 Bandwidth Growth Cost

It is proving very difficult to make a business case for mass market deployment of the FTTP solutions described above. The problem is not simply the cost of the FTTP access solution but also the backhaul/metro and core network build that will be needed to support it. These costs are often ignored in comparative analyses of access solutions because usually, they are considered to be a common cost and indeed, for the FTTP architectures discussed above, that is generally the case. However, considering FTTP to be an access only problem is missing the point because it is the problem of the total end-to-end cost growing beyond any potential revenue growth that ultimately will limit FTTP deployment. The problem of the cost of meeting bandwidth growth exceeding

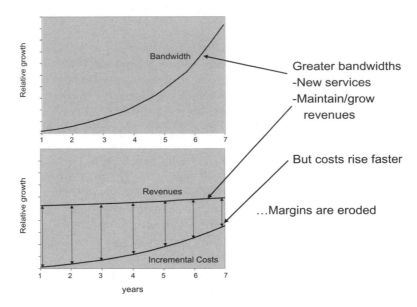

Figure 8.5 High bandwidth growth – eroding margins

revenue growth is described in more detail in Payne and Davey (2002), and arises from the simple macroeconomic observation that revenue growth in the telecommunications sector has been relatively static for the past few years and that the price of bandwidth has declined only in line with equipment price declines (the bubble around the turn of the millennium has distorted these figures, but the average trends have been fairly constant). When these two observations are combined in a simple analysis involving extrapolations of future broadband service demands, then the problem of revenues being outstripped by the cost of providing the network capacity becomes apparent; this is illustrated in Figure 8.5. The problem seems not to be solvable by normal equipment price declines; the potential future bandwidth growth is too large in most broadband service scenarios utilising FTTP capabilities.

All the FTTP architectures described in the previous section rely on equipment price decline in order to be viable for mass roll-out because they do not fundamentally change network economics; in particular, they keep the backhaul and access networks separate and require multiplexing and aggregation electronics at the local exchange site. The cost of the local exchange and the corresponding backhaul networks can be as large as the access network, when high bandwidth and low contention service requirements are to be delivered.

Of course it could be that the high bandwidth service demand does not arise, either because it is not affordable or the network isn't built to be capable of supporting such services. If either is the case, then optical access networks will not be required and the argument becomes academic. However, if the services are to be provided and FTTP is required, then the only conclusion is if price decline of network equipment is not sufficient, the only other option is to find architectures that can remove the equipment from the network.

8.4.2 A New Architectural Approach – Long-reach Access

The conclusion the author and other members of the BT optical research team came to several years ago was that a new architectural approach was the only way to overcome the cost of bandwidth growth problem (Payne and Davey, 2002). As a result, we have developed the idea of the long-reach access solution that eliminates the separate backhaul network and also the electronic aggregation node at the local exchange or central office. The basic architecture is shown in Figure 8.6; it shows a simple single wavelength system, which is basically an amplified PON using erbium doped fibre amplifiers (EDFAs) to boost the optical signal, enabling 100 km reach, up to 1000 way split (to share the backhaul fibre as much as possible and also to get maximum statistical gain) and a line speed of 10 Gb/s. The 10 Gb/s line rate provides a sustained or average rate of 10 Mb/s per end user, but more importantly enables end users to burst for short durations up to Gb/s speeds, ultimately limited by the port speed of the end user ONT (also called an optical network unit). This is the burst-mode of operation described earlier and allows large files to be transferred very quickly; even feature-length films could be transferred in a few 10 s of seconds at such speeds.

This network architecture fundamentally changes the network economics by enabling bypass and eventual elimination of local exchanges/central offices and also elimination of the separate backhaul or metro network. The long-reach PON head-end optical line termination (OLT) is situated in the metro/core edge node and replaces the equipment

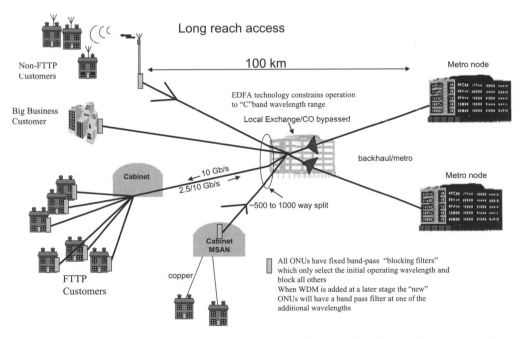

Figure 8.6 Basic long-reach access architecture – could reduce UK network to ~100 nodes. MSAN, multi-service access nodes. ONU, optical network unit

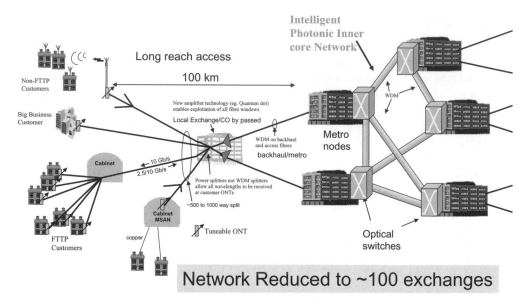

Figure 8.7 Long-reach access – long term vision with flexible wavelength assignment

that would terminate transmission systems in a conventional network. The architecture therefore provides major reductions in capital equipment by removing the need for multiplexing, switching and transmission equipment at the local exchange site and also provides massive reductions in operational expenditure by reducing a network the size of the UK from about 5600 electronic switching nodes to about 100.

8.4.3 Future Evolution

A further stage of evolution is shown in Figure 8.7. Here, WDM has been added to increase both the system capacity and to provide flexibility in the wavelength domain. Flexible wavelength assignment coupled with sharing bandwidth within each wavelength using time multiplexing protocols (packet or time slot based) enables the ultimate form of bandwidth management to end users and communications providers. Also shown for completeness is the optically switched inner core, which, together with the long-reach access, could reduce a UK-sized network to only ~100 large nodes.

The team at BT experimentally verified that from an optical power budget and transmission perspective, such a long-reach access system is technically viable (Nesset *et al.*, 2005), and WDM options are being investigated in the European Union (EU) collaborative project PIEMAN.[1] A prototype version of a long-reach PON was demonstrated by Siemens in March 2007; this work was carried out as part of the MUSE[2] (MUltiService Everywhere) EU collaborative project and demonstrated a long-reach PON with 512 way split and an operating protocol working at 10 Gb/s downstream

---

[1] www.ist-pieman.org/.
[2] www.ist-muse.org/.

and 2.5 Gb/s upstream. The protocol is also capable of 10 Gb/s upstream when the 10 Gb/s Burst-mode receiver for the head-end equipment is available (Rastovits-Weich *et al.*, 2007).

There are a number of technical and design issues still to be resolved for a viable long-reach access solution to be developed into a standard product ready for deployment and realistically, it is several years away. An interim solution that paves the way for this architectural approach is 'extended reach GPON' (Nesset *et al.*, 2006) As mentioned above the GPON protocol can support up to 128 way split and 60 km range, but the standard optical components cannot realise this capability in a practical system. By adopting the ideas from the long-reach access vision, an amplifier module can be added to GPON to realise the full protocol capability while keeping the standard GPON optical component specification.

Amplified GPON (A-GPON) can offer a short term expedient to long-reach access, enabling exchange bypass of a large proportion of the local exchanges and elimination of the backhaul transport network where it is deployed. The fibre architecture is compatible with long-reach access and could be upgraded at a later date when the long-reach access system becomes available for deployment and the service demands and business environment make it worthwhile.

8.5 The Impact of FTTP on the Core Network

8.5.1 Core Traffic Growth

From the example results shown in the discussion on bandwidth growth in Section 8.2.3, it can be expected that with the advent of FTTPs, average sustained bandwidths per end user in busy periods could easily approach 10 Mb/s. This level of bandwidth growth is unprecedented and is a huge increase in the levels of traffic networks will be required to cope with. To get some feel for the impact of such growth, it is worth considering some very simple order of magnitude numbers for the UK network:

If we assume that 50% of end user premises will take up and use an FTTP offering, then we could expect:

- 10 million end users;
- 10 Mb/s per end user;
- total sustained busy period traffic = 100 Tb/s;
- assume 100 km reach access – long-reach passive optical network (LR-PON) architecture = ~100 core nodes;
- average traffic per node = 1 Tb/s;
- average traffic between each node pair = 10 Gb/s (assuming little turnaround traffic, no protection, no bandwidth contention, etc.).

If traffic levels are of order 10 Gb/s between all node pairs, then this is equivalent to assigning one optical wavelength between all node pairs which in turn means no need for grooming/intermediate electronics. That is, the core network could be a flat, single layer all optical network and the long-reach access and metro/core node

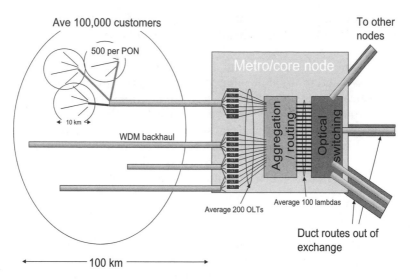

Figure 8.8   Metro/core node architecture

structure could be similar to that shown in Figure 8.8. This architecture, based on the long-reach access architecture discussed above, reduces the UK network from one with around 5600 exchange locations containing electronic equipment to one with only of order 100 electronic switching/routing nodes. It should also be pointed out that the dimensioning of the 100 core nodes in terms of total equipment count is no greater than that required for the core nodes in the 5600 exchange architecture. In fact, the use of an all optical core network will considerably reduce the amount of electronics required within the core nodes. The multiplexing and traffic grooming functions carried out by local exchange electronics is now carried out within the protocol operating on the long-reach access networks so that this functionality is not added to the core node equipment requirements.

8.5.2 Node Bypass

The reduction in core node electronics can be seen by considering the traffic paths required across a 100 node core network. All nodes will need at least one wavelength to all other nodes. These optical paths will traverse the network and in doing so will pass through intermediate nodes; the distribution of the number of nodes passed through by these optical paths is shown Figure 8.9. In today's network, each optical path would be terminated at each node it passes through requiring opto-electronic conversion on the ingress and egress ports for every wavelength. The traffic carried on each wavelength would then be switched/ routed through to the egress port requiring the switch capacity to cope with all the through traffic as well as the traffic destined for that particular node. Studies of the UK network show that as much as 70%–80% of the traffic entering a typical core node will be through traffic. If this traffic remains in the optical domain while passing through the node, then there are enormous savings in opto-electronic port cards/transponders and also significant reductions in switch/

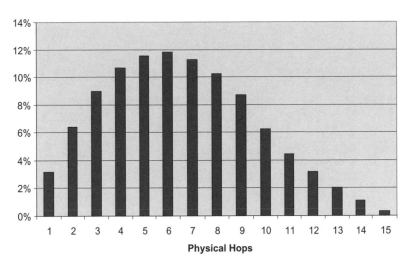

Figure 8.9 Number of nodes passed through

router capacity required. To do this economically, what is required are cost-effective optical switches added as an optical layer at each metro/core node.

8.5.3 Flat v Hierarchical Core Network

This optical layer effectively creates a flat rather than hierarchical core network. The flat architecture avoids the need for the opto-electronic conversions at layer boundaries as the transmission paths traverse the network; it also reduces the size of the routers required by not electronically handling the through traffic. The hierarchical architecture, however, allows for sub-wavelength grooming whereby traffic from several nodes could be multiplexed on to a shared wavelength for onward transmission to a destination node.

From this simple qualitative argument, it can be seen that for low traffic levels, a hierarchical core would use optical transmission capacity more efficiently, while for high traffic levels, the flat architecture makes more efficient use of core node resources. It can be expected, therefore, that from a cost perspective, a transition will occur from hierarchical to flat networks as traffic levels rise, and to minimise cost, networks will need to evolve from a hierarchical architecture to a flat all optical architecture. This transition is illustrated in Figure 8.10 and shows that at traffic levels around 40 Tb/s, the flat all optical core architecture becomes the optimal solution. The flat architecture is expensive for low traffic levels because the flat core requires large numbers of wavelengths at each node to simply provide the wavelength mesh connectivity required and requires high upfront expenditure. At low traffic levels, the wavelengths are inefficiently filled, but traffic can grow to high levels without further expenditure. The hierarchical network, however, can efficiently fill the wavelengths by use of sub-

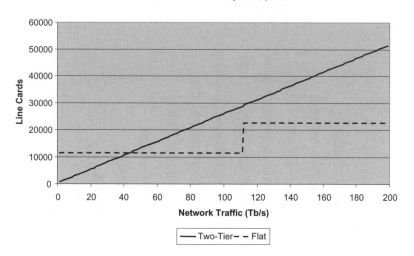

Figure 8.10   Economic crossover for flat architecture

wavelength grooming and therefore requires significantly fewer optical wavelengths in the core to provide the required connectivity. However, as traffic grows, the equipment required grows in proportion, and therefore crosses the flat core curve at some traffic level. The parameter used for cost in Figure 8.10 is line card count. Line cards account for a large proportion of the cost of core nodes, and line card count is therefore a useful comparator for architectural cost studies.

Today, traffic levels in the UK network are around 2–3 Tb/s, well below the 40 Tb/s transition point, and the hierarchical architecture makes eminent sense; it is therefore still used for BT's 21CN network. The challenge for the future evolution and growth of the network is to find graceful ways for the transition from the electronics based, hierarchical network of the 1st generation 21CN network to the flat all optical architecture that will be required for the high traffic levels arising if FTTP is deployed for the mass market.

8.5.4 Future Core Node Architecture

The simple analyses used above made the gross assumption that all core nodes are equal; of course in the real network, this is not the case, and in practice, there will be a considerable range of core node sizes. The larger core nodes will have several 10s of fibre links connected to them even with 80 channel dense wavelength division multiplexing (DWDM) systems. Many of these fibres will have all the wavelengths carried as through wavelengths to other nodes. The architecture of the metro/core node may therefore need two layers of optical switching; one switching at the fibre level and one at the wavelength level. This latter switch could be a wavelength selective switch (WSS). The architecture of the node would then look something like that shown in Figure 8.11.

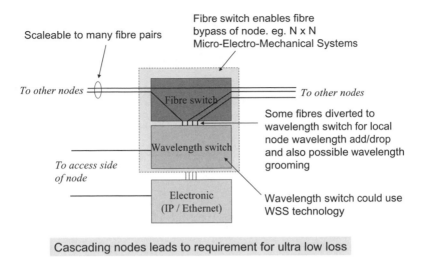

Figure 8.11   Future core node architecture

The optimisation of such architectures, getting the right balance between fibre switching, wavelength switching/multiplexing and electronic switching/routing/multi-plexing functions is still an area of active research as are the methods and technologies that can enable graceful growth in all layers of the node hierarchy such that the electronic layer grows in proportion to the traffic terminating on the node while the wavelength switching layer grows in line with through traffic growth. This optical layer should also be capable of gracefully growing the fibre switch layer rather than the wavelengths layer as further growth occurs and the traffic on whole fibres passes through the node.

Of particular relevance to this all optical architecture is the balance between wavelength multiplexing and higher transmission speeds. High transmission speeds can mean lower wavelength counts and fewer transponders at terminating nodes. However, getting efficient fill on higher granularity wavelengths may require sub-wavelength grooming which gets away from the advantages of the flat optical core network. More wavelengths give more optical switching flexibility and possibly more graceful growth but at the expense of more line equipment. An additional advantage may be greater resilience and therefore networks with greater overall availability.

It may be that higher speed systems (40 Gb/s or 100 Gb/s) are used selectively on the highest capacity links with 10 Gb/s being used more ubiquitously on the larger number of lower capacity routes. The actual choice will depend as much on economics and resilience as technical capability of the transmission equipment and is currently an active research topic.

8.6 Timeline for All Optical Networking

Most of the technologies required for the all optical network vision exist today. There are outstanding technical problems associated with the access network particularly the

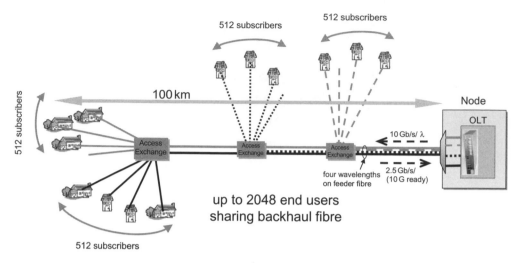

Figure 8.12   Prototype LR-PON system – Nokia–Siemens

low-cost high-speed transceivers that will be required for the LR-PON solution and the 10 Gb/s burst-mode receiver for the head-end OLT. The A-GPON solution is being developed, and standards for a generic GPON extender box are in progress within FSAN[3] and the ITU[4] with ratification expected in early 2008.

An important component for future evolution of optical access systems is the deployment, at initial installation, of a blocking filter in the end user optical termination (ONT). This filter has now been agreed and standardised in FSAN and the ITU, so given a fair wind for the generic extender box, it can be expected that commercial and standards compliant A-GPON solutions could be available before 2010 with proprietary solutions available before then.

The 10 Gb/s LR-PON solution may arrive in stages. The first stage may be a 10 Gb/s downstream and 2.5 Gb/s upstream solution to postpone the requirements for a 10 Gb/s burst-mode OLT receiver and end user ONT transceiver. Indeed as mentioned earlier, Siemens through the EU MUSE project have already demonstrated such a system, as shown in Figure 8.12.

When the 10 Gb/s ONT transceivers and OLT burst-mode receivers are available, the LR-PON could gracefully transition to the 10 Gb/s symmetrical system. The use of blocking filters for the LR-PON system is also going to be required at day one installation, and some detailed and careful work is required on the wavelength plan for the evolution roadmap for LR-PON future developments. The long term vision would be to use tuneable ONTs exploiting the DWDM ITU grids initially in the optical C-band (1525–1565 nm) where EDFAs can be used for the first generation systems. The evolution strategy for LR-PON should include a path for continuous evolution for exploitation of new technologies as they emerge in the future. Careful design of

---

[3] www.fsanweb.org/default.asp.
[4] www.itu.int/net/home/index.aspx.

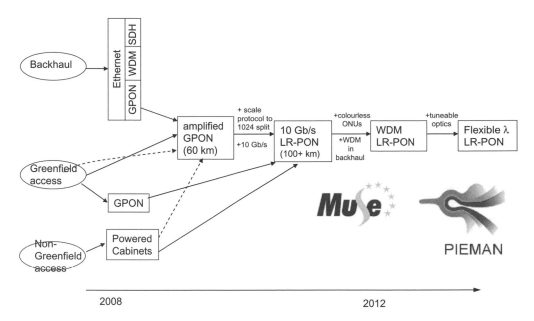

Figure 8.13 Possible timeline for optical networks (SDH, Synchronous Digital Hierarchy) (Reproduced by permission of BT)

the physical infrastructure coupled with the right wavelength and evolution plans can ensure this. However, it would also be very easy to design these networks to become legacy dead ends as with the copper network today. This can and must be avoided.

A possible evolutionary timeline from standard GPON through to the fully flexible LR-PON with optical core network is shown in Figure 8.13. The times shown are possible timescales given the development already taken place; the actual timeline will depend on a multitude of political and economic circumstances, not the least of which is simply the political will to carry out such a dramatic modernisation of the network and also the demand for the services that could then be delivered over such a network.

The optical core network depends on the commercial realisation of optical switching systems. Optical switching technology development collapsed after the millennium bubble and much of the research work stopped. However, a few of the small start-up companies continued and optical switching systems are now available that are fit for purpose for carrier class operation. One of the major areas that needed development was not so much the optical switching component technology but the management and operational support systems that operators must have if they are to deploy such technology throughout their core networks. These management and support systems have now reached a sufficient level of maturity to be deployable in carrier networks.

There are still developments required of the optical switching component technologies. WSS are still quite small and suffer quite high optical loss. Structures that can grow gracefully and be economically applied to a wide range of node sizes still need further development. Similarly, optical path switches need structures that allow grace-ful, economic growth so that upfront investment can be minimised and expenditure kept more in line with traffic growth.

8.7 Conclusion

The demand for bandwidth is unlikely to diminish in the future. New broadband ser-vices with richer content containing more video, including higher definition material, will demand huge increases in the bandwidth networks will need to support. Tradi-tional networks will struggle to scale and remain economically viable if this bandwidth growth is realised. The access network can evolve to exploit FTTP and deliver huge increases in bandwidth. However, the hierarchical nature of today's networks with separate transmission systems in the backhaul/metro and core networks intercon-nected via large numbers of electronic aggregation and switching nodes does not scale economically. Models of potential service and bandwidth growth coupled with cost models of conventional networks suggest that the cost of providing the bandwidth growth that could be demanded will exceed any prospective revenue growth.

The fundamental problem is that the price of the electronic equipment will not be able to decline fast enough. To overcome this problem, it is proposed that all optical networks are used to eliminate large swathes of the network reducing the number of layers and nodes, massively reducing the amount of electronics required and avoiding this cost of electronic equipment. This can realise a network that not only requires vastly less capital investment but also can enable significant reductions in operational costs via the elimination of a large proportion of the equipment and nodes required.

The architecture that will win out will use long-reach optical networks in the access that bypass local exchanges/central offices avoiding the need for separate backhaul/metro transmission systems. The traffic from these large, long-reach access networks will terminate onto a small number of core nodes which will in turn be interconnected by a flat optical core network. This network approach will use less power, be lower cost to install and operate, and can grow the total capacity of the network by several orders of magnitude compared with today's architectures.

Acknowledgements

The author would like to thank the optical research team in BT Design at Adastral Park, Martlesham Heath for their contribution to and support of the vision architec-ture described in this chapter. In particular, he would like to thank Russell Davey and Andrew Lord for their direct contributions to the material used in this chapter.

References

Halpern, J., Garceau, G., and Thomas, S. 2004 Fibre: Revolutionizing the Bell's Telecoms Networks, http://www.telcordia.com/products/fttp/bernstein_report.html
Lehr, W., Gillett, S., and Sirbu, A. 2005 Measuring Broadband's Economic Impact, *Broadband properties*, December 2005, http://www.broadbandproperties.com/

Nesset, D., Davey, R., Shea, D., Kirkpatrick, P., Shang, S., Lobel, M., and Christensen, B. 2005 10 Gbit/s Bidirectional Transmission in 1024-way Split, 110 km Reach, PON System using Commercial Transceiver Modules, Super FEC and EDC, *Proceedings of ECOC 2005*.

Nesset, D., Payne, D., Davey, R., and Gilfedder, T. 2006 Demonstration of Enhanced Reach and Split of a GPON System Using Semiconductor Optical Amplifiers, *Proceedings of ECOC 2006*.

Payne, D. and Davey, R. 2002 The future of fibre access systems, *BT Technology Journal* Vol. 20 No. 4 October 2002.

Payne, D., Davey, R., Faulkner, D. and Hornung, S. 2006 Optical Networks for the Broadband Future in Lin Chinlon (editor) *Broadband Optical Access Networks and Fiber-to-the-Home: Systems Technologies and Deployment Strategies*, Chapter 8.

Prestowitz, C. 2006 America's Technology Future at Risk: Broadband and Investment Strategies to Refire Innovation, *Economic Strategy Institute*, March 2006, http://www.econstrat.org/.

Rasztovits-Wiech, M., Stadler, S., Gianordoli, K., and Kloppe, K. 2007 Is a 10/2.5 Gb/s Extra-Long PON far from Reality, *ICTON 2007*, Rome, 1–5 July 2007.

9

End-to-End Service Level Agreements for Complex ICT Solutions

John Wittgreffe, Mark Dames, Jon Clark and James McDonald

9.1 Introduction

Today, corporate organisations today are almost completely dependent on the availability and responsiveness of their networked applications. If a networked application slows down, so too does the organisation; if an application fails, work stops and the company loses money.

The operational environment, however, is increasingly complex. A medium-sized to large organisation may use more than a hundred business applications, which can range from Microsoft Office, to SAP enterprise solutions, to applications built in-house and unknown to the outside world. These applications can run on complex application software stacks and can operate over local area networks (LANs), over different backhaul networks, across IP-VPN (virtual private network using internet protocol for packet routing) wide area networks (WANs), WAN interconnects, data centres, servers and virtual servers. Each and every network and IT component contributes to the overall perceived performance (Wittgreffe and Dames, 2004) (Figure 9.1).

Whilst service providers may be well versed in offering service level agreements (SLAs) focused on the availability and performance of each component individually, the end customer is more concerned with the performance they experience from the applications in use . . . 'is the delay too long?', 'is the quality of the picture good enough?', 'does my web page load correctly?'. Basic SLA metrics, such as ensuring the availability of a network access link, are far removed from this user view. Instead, this requires a shift of service provision from focussing on the underlying network

ICT Futures: Delivering Pervasive, Real-time and Secure Services
Edited by Paul Warren, John Davies and David Brown
© 2008 John Wiley & Sons, Ltd

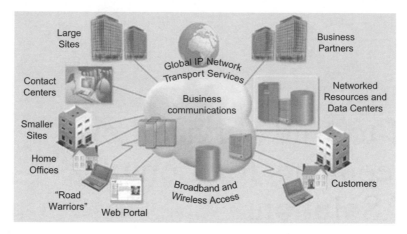

Figure 9.1   The service perceived by the end user spans potentially multiple ICT components 'end-to-end' across corporate and service provider infrastructure

products to a focus on achieving targets for application performance end-to-end across all ICT infrastructure. The related SLAs then reflect whether user sessions are 'good' or 'bad' rather than the individual product performance; the maximum extent of allowed 'bad' performance then becomes a contractual obligation.

Whilst this is an easy concept to understand, the complexities of providing such an SLA are potentially immense – spanning the interrelationships between all underlying components that contribute to the perceived performance *and* potentially spanning the contributions of multiple different service providers.

In this chapter, we begin to address pragmatically the means to deliver an application-level SLA. Firstly, we outline the potential formats of application-level SLAs and the associated issues. We then focus on the near term future and offer a pragmatic five-step technology roadmap for meeting application-level service level guarantees (SLGs) within a corporate ICT environment, bringing the performance of networked services in line with business objectives and priorities. We also propose a longer term solution based on comprehensive SLA negotiations between ICT components.

9.2 Broadening the SLA Considerations

SLAs can take almost any form, but at their most basic level, they are a contractual agreement between the service provider and their client as to the expected level of 'service'. If this level is not met, there may be punitive consequences such as the issue of refunds or service credits. In the case of punitive compensation, the SLA is sometimes referred to as an SLG, although within this chapter, we do not distinguish. We refer only to an SLA, which – dependent on the contract – may or may not include punitive compensation dependent on the contract.

In terms of measures of target service levels, the typical contracted measures today include % uptime, the average response time to a fault and the average time to fix a fault. These are complemented by specialised ICT performance measures specific to

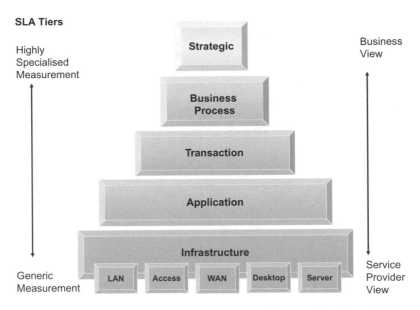

Figure 9.2 SLA tiers of measurement, moving from a service provider view of their own products to the customer's own business view

the domains, such as bandwidth provided = bandwidth promised, Multiprotocol Label Switching (MPLS) classes of service (CoS), network latency, network jitter, server resources, etc. SLAs may include other parameters such as security obligations. In major client solutions, a master SLA may bring together common measures across multiple related products that make up an agreed service, even across multiple service providers who contribute to the solution, for example, the average time to fix a fault in a WAN may span multiple network access providers.

Whilst these infrastructure-level measures are very important, SLAs are evolving to application-level SLAs today, and in future even beyond these, as summarised in the SLA Tiers diagram shown in Figure 9.2.

The bottom tier is the Infrastructure tier where SLAs are a standard way of doing business. The SLAs are infrastructure, rather than application-focused.

The application tier focuses on the performance of individual applications as perceived by the end user. This tier is characterised by overall application quality measures, derived from measuring application flows (flows of information associated with a particular application session) or client/server responses.

The transactional tier focuses on a particular task that is entirely delivered by the ICT such as a web search or confirming an order. The transactional SLA is characterised by an overall response time to that request, and may constitute a number of separate but related application flows and client/server responses.

The business process tier focuses on an end-to-end process in the client's own business, which may involve some tasks delivered by the ICT, but also manual tasks with potentially no involvement from the ICT at all. The measurements are characterised by key quality indicators (KQIs) and key performance indicators (KPIs) – the classic case in the telco sector is contact centre outsourcing with metrics such as time to answer call, call clearing rate, customer satisfaction, etc.

The strategy tier focuses on performance against the customer's own strategic scorecard, such as revenue, bottom line, profit, customer satisfaction, international growth, etc. Whilst it would be exceptional for a service provider to underpin a client's performance against their strategic scorecard with an SLA, it is still important, as far as possible, to understand the relationships between SLA performance and impact on the client's top-level objectives (Vitantonio *et al.*, 2006).

The application-level and transactional-level SLAs represent the primary foci of service provider developments. These two areas focus on the end-to-end performance experienced by each user, individually, and provide a bridge between the infrastructure view and the process and strategic views of the client's business. The application-level and transactional-level SLA are considered in more detail below.

9.3 Application-level SLA

Application-level SLAs focus on the performance of individual applications end-to-end across the entire infrastructure, according to how the end user perceives the application. Whilst the focus is on an entire quality score for an application, this must be derived from tangible and real measures from the underlying infrastructure: the end users cannot continually be asked how they are perceiving performance. Firstly, we consider the underlying measures, then we consider the overall quality score.

9.3.1 Measures

The tangible measures usually relate to application sessions and their performance in different parts of the infrastructure, for example:

- minimum bandwidth for a session of application X, end-to-end across the entire network path;
- maximum latency for a session of application X, end-to-end across the entire network path;
- maximum jitter for a session of application X, across the entire network path;
- maximum round trip delay for a session of application X;
- server response time to the application request;
- maximum allowed packet loss for a session of application X;
- expected response confirmed – that the information returned from a query to the application is as expected.

Specific measurement targets can then be set up for each application of interest, describing what makes a 'good' session for the end user; an example is in Figure 9.3.

In the example in Figure 9.3, performance metrics are set for each application: in line 1 a transactional function on 'SAP' requires 50 kbit bandwidth and delay of 150 ms max. Note that the quality metric may need to be specific to the application function, not the application in general. For example, a 'buy now' button on a web site may

Business Application Service Level Guarantees						
APPLICATION	PRIORITY	TARGET BANDWIDTH (kpbs)	LOSS (%)	DELAY (ms)	JITTER (ms)	Amber, Red (time outside SLG)
VoIP	1	32	0	100	50	Av. over 2 sec, Av. over 10 sec
SAP	=2	50	2	150	n/a	Av. over 10 sec, Av. over 20 sec
ORACLE	=2	20	3	200	n/a	Av. over 10 sec, Av. over 20 sec
CEO Video Broadcast	4	1500	1	200	80	Av. over 2 sec, Av. over 10 sec
Video Conference	5	500	1	200	80	Av. over 5 sec, Av. over 20 sec
Web	6	30	3	500	n/a	Av. over 1 min, Av. over 10 min
EMAIL	7	25	5	n/a	n/a	Av. over 1 min, Av. over 10 min
FTP	8	25	5	n/a	n/a	Av. over 1 min, Av. over 10 min
Unclassified	9	10	5	1000	n/a	Av. over 2 min, Av. over 20 min

Figure 9.3   An example of application-level performance targets (FTP, file transfer protocol)

have a critical delay dependency, but the customer may be less concerned about the delay for 'company background pages' from what is perceived by the end user as the same application (and in fact running on the same server via the same http port!). These specific measures can then be rolled up into an overall quality score.

9.3.2 Overall Quality Scoring

To simplify the management of application-level SLAs, these individual application metrics are best combined into an overall quality measure of performance for the application, such as mean opinion scores (MOS) offered by Psytechnics[1] for voice or video, or a custom combination of the metrics or flows for a particular web application.

This overall quality measure has more meaning for the end customer and can be expressed very simply for the purposes of a contract in terms red/amber/green/black for each application session, for example,

- green – good performance as expected;
- amber – below expectation but can continue work;
- red – unacceptable, some aspects no longer work;
- black – simply not working at all.

_____
[1] http://www.psytechnics.co.uk/.

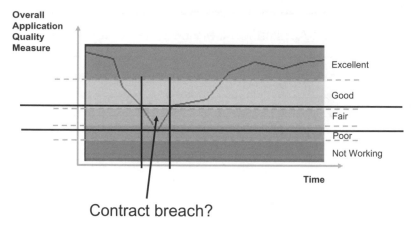

Figure 9.4   An example overall quality scheme showing a possible period of contract breach

An example of a general application quality scoring scheme is provided by the APDEX[2] organization, which aims to provide standardised interpretations of application performance. Their work so far has largely been focused on client/server delay; however, APDEX is moving into wider application quality scores.

With the simplification to an application quality score, the general principles of SLAs can be applied to formulate a simple commercial offer in the form of an SLA focused on the application or the transaction, based in part on the periods of time for which the application is within different colour categories, for example,

* uptime;
* time within green condition;
* time within amber condition;
* time within red condition;
* time within black condition;
* response time to fault;
* fix time for fault;
* % of users experiencing each category;
* etc.

Any punitive compensation is then based on the time for which the application is outside the quality target. This 'level' of time outside performance threshold again very much depends on the application. A business TV broadcast is very sensitive to quality problems, where a 0.5 s glitch at a critical moment of a speech is serious. By contrast, a 0.5 s glitch in a large file transfer is likely to have little consequence (Figure 9.4).

---

[2] http://www.apdex.org/.

9.3.3 Reverse SLA Parameters

One of the unique features of an application-level SLA is the potential need to specify 'reverse SLA' parameters. Some applications which are known as noncritical may not need assurance of good performance, but conversely may require assurance that the applications will never use more than a certain percentage or amount of resources, in order to prevent escalating infrastructure costs for noncritical applications. A classic case is for Internet use – not blocking Internet use but deliberately limiting the total bandwidth given to any given Internet http session, or the total bandwidth for Internet from any office site. Whilst this may only be critical at times of contention, it may be important for cost control.

9.3.4 Exceptions

As any IT expert will comment, the performance of an application isn't just about how good the infrastructure is – the way in which the application is used affects performance. For example, a web site may perform fine one day, but the next suffers a 'rush' in response to an event and performance drops even with the same infrastructure. Reasonable limits must thus be defined beyond which performance degradation is not a contractual obligation. Otherwise, a client could arrange a 'denial of service' attack upon their own services at the end of the month in order to rake in rebates and avoid payment!

In addition, the service provider is unlikely to underwrite performance problems that are completely outside their domain of control. An example is a performance problem due to a bug in an application the client has written themselves and over which the service provider has no control. This does not mean that an application-level SLA cannot be offered in this circumstance, it simply means that if the domain of responsibility is found not to be within the service provider's control, there are no punitive consequences on the service provider. A solution of the type described later in this document can maintain real-time views of the performance of ICT components and rapidly calculate the domain of responsibility for any application-level SLA breach.

9.3.5 Business Rules

Not every application is important, and application priorities vary from customer to customer. Moreover, different applications may be more important at some locations than at others. An application-level SLA, therefore, has to be managed not just according to performance metrics and quality targets, but also according to business rules, specific to each customer, describing what is important, where and when:

- which applications are key and are to be assured under an SLA, and which are not;
- the priority or criticality of the business application to the customer;
- variations in policy per site (e.g. the headquaters may have better SLA than branches for the same applications);

- variations in policy per type of user;
- variations in policy at different times (by day local time an application may be critical, but at night unimportant).

9.4 Transactional-level SLA

In some cases, the application-level quality measure maps 1:1 onto the user experience; however, in some cases, it does not. For an application flow using user datagram protocol (UDP) protocol such as video, there is a clear 1:1 mapping between the MOS score for the video flow and the perception of the end user. For a rich web page, however, there may not be a clear mapping to a single flow or client/server response, where the construction of the response may include multiple separate client/server calls to different servers at different locations. This may include, for example, text from a catalogue system, stock levels from a stock control system, a video window explaining the product from a video streaming system, etc. In such complex cases, the next level of SLA is required, the transactional-level SLA.

In our definition, a transaction is simply the user's own perception of the use of an application for one specific purpose, for example, a click on 'order now' and receiving confirmation, or performing a web search and receiving the returned page. In many cases, this 'transaction' will be underpinned by multiple different application calls, as shown in the example in Figure 9.5.

In the example in Figure 9.5, the user perception is a combination of the video flow, the catalogue information and the stock information. Interestingly, if the video is poor

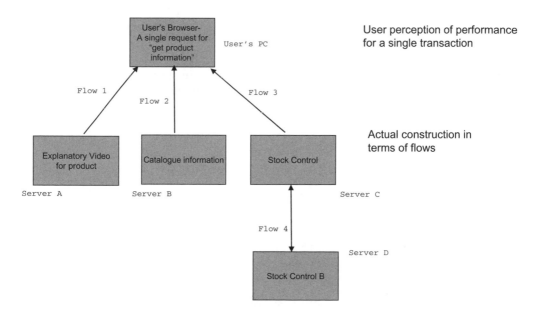

Figure 9.5 A transaction may span multiple application-level interactions, requiring a 'transaction-level SLA'

quality, the perception may be worse than if it is not working at all; if it is not working, the impact is simply that a section of the screen is blank; the user can still obtain the catalogue and stock information and can still progress with the order. If the stock and catalogue system is performing slowly, this may be slightly frustrating but not as bad as poor video, whilst if the stock and catalogue information failed altogether, this would be critical.

From this example, it is clear that the implementation of the transactional-level SLA can be an order of magnitude more complex than an application-level SLA, requiring an understanding of the multiple application-level interactions behind the transaction, their individual performance contributions, performance weightings and the domains of ownership. This is usually highly specific to each customer.

It is therefore always worth distinguishing between the two different levels of end-to-end SLA, application level (based on a single application message flow) and transactional level (composites of a number of flows). It is essential to understand the risks and consequences of attempting to implement a transactional level, rather than an application-level SLA, especially in cases where the transaction cannot be described in its entirety by a small number of well-understood application flows and client/server responses.

9.5 Requirements on Technology

Clearly, achieving these higher level SLAs presents significant risks to a service provider, *unless* additional technology is available to manage application performance to the required levels within the infrastructure, and to identify any exceptional circumstances. Whilst the proposition of an application-level SLA is compelling to the customer, for the service provider, this means extra capability both in terms of equipment within the network, operational complexity, and associated training of sales and field staff.

The primary requirements on technology are

- The ability to audit a customer's infrastructure and applications, to establish the application flow trends, understand which are key and which are background applications, what constitutes 'good' performance for applications, areas that are out of domain, and the limitations across the end-to-end network and IT paths. This base-level information is required, usually in close consultation with the customer, before any application-level SLA terms can be defined.
- The ability to set application-level quality measures and targets, and to monitor and report on performance against those targets as experienced by all users at all locations. This includes both reporting in real time and historically.
- The ability to vary the targets for different types of users/sites according to the client's own business policy.
- The ability to identify in real time the application, application function, user, site and/or any combination of identifying parameters against which quality targets are defined (e.g. high quality is only needed for the 'order now' button on Siebel for users from call centre X).

- The ability to control as far as possible, dynamically and in real time, the underlying network and IT resources to ensure that application-level SLAs are met at times of contention/high demand.
- The ability to perform root cause analysis, not just for fault resolution, but to clearly distinguish in-domain from out-of-domain in terms of SLA consequences.
- The ability to identify and effect punitive measures in response to SLA breaches, for example, the issue of service credits.

Additionally (and optionally):

- The ability to provision more resources on demand (potentially for a premium charge) to meet exceptional out of contract needs that are not met by optimising within the existing provisioning.

This is by no means an exhaustive list but covers the key features arising from the preceding discussion.

It is evident that the implementation of the application-level SLA can quickly become complex; however there are pragmatic ways to address this, exemplified by our example solution below.

9.6 An Example Solution

By combining current state-of-the-art technologies, it is possible to create a system with many of the features required for application-level SLAs. This example solution focuses on delivering such SLAs, using a combination of existing BT products and technologies under development at BT Laboratories.

Figure 9.6 shows five primary technology steps taking us from the current baseline SLA focused on IP-VPNs through to a comprehensive closed-loop system incorporating policy-based, on-demand provisioning. Relevant products and developments from BT are also named as examples.

- Step1: MPLS IP-VPN with CoS – the key baseline product enabling differentiation of business-critical and noncritical traffic across the IP-VPN WAN (Carter, 2005), e.g. BT's MPLS portfolio. This allows basic SLAs based on overall network performance for differentiated application traffic, with the core network only.
- Step2: Application Discovery and Reporting – the ability to audit the capabilities of IT and network infrastructure, discover applications running over a corporate network and report against application-level performance targets. A dashboard shows the performance of the applications and ICT infrastructure in the context of the customer's SLA. Trends in performance allow prediction of future resource needs, e.g. BT's Application Assured Infrastructure services (Dann et al., 2005).
- Step 3: Application-level Optimisation and Alerting – monitor each application session in *real time* and *optimise* application flows over the end-to-end network. Ensure critical business applications always get resources they need, and alert in

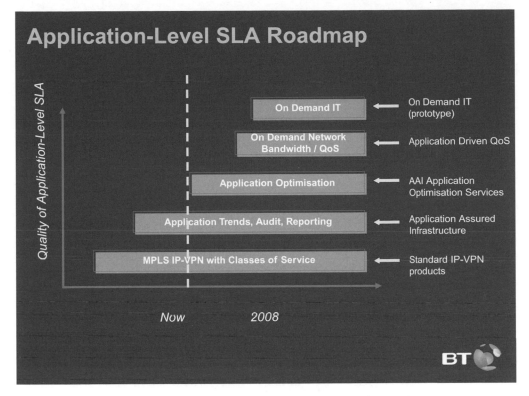

Figure 9.6   Roadmap of technology for application-level SLAs (Reproduced by permission of BT)

real time on 'amber' or 'red' conditions, e.g. BT's Application Optimisation Service (BT infonet). Alert on the source domain of any problem e.g. LAN, WAN, or server.

- Step 4: Application Driven QoS – Where optimisation is not enough, temporarily provide additional bandwidth and/or improved QoS in real time according to policy (Dames *et al.*, 2005) and the use of quotas/spend caps (e.g. BT's 21st Century Networks Application Driven QoS Capability).
- Step 5: On-demand IT – the combination of on-demand network, processing, and storage resources, dynamically responding to application needs according to the driving SLAs (e.g. BT's Virtual Service Platforms development under the 21C programme).

There are of course much wider considerations for service management, including ITIL compliance of the operational systems and processes, and ITIL[3] trained staff, and the means by which multiple service providers are linked together for processes such as fault resolution. These considerations are beyond the scope of this chapter, but are presented in detail in Wittgreffe *et al.* (2006).

---

[3] http://www.itil.org.uk.

9.7 Longer Term – The Comprehensive Approach

Whilst the roadmap shown above is a pragmatic way to deliver many of the features of transactional-level SLAs and application-level SLAs longer term, a more comprehensive approach is being determined by international collaborations such as the European NESSI[4] programme and the Telemanagement Forum[5], that can be applied from very high business-level SLAs through to the smallest performance contributor. This requires each SLA 'customer' to agree and define terms with each SLA 'contributor' within a service oriented architecture environment. This can then be policed and managed to make sure that each contributor meets the agreed objectives at run-time. Chapter 8 introduces this more general approach.

Further, ongoing R&D work is enabling us to put artificial intelligence into the support of SLAs at the higher tiers. BT labs already has a 'real-time business intelligence' facility (Azvine et al., 2007), that can deduce through correlation techniques the interdependencies between process performance, application performance, and infrastructure performance. This is in turn used to help configure ICT to meet process KQIs and KPIs, always remembering that process performance is not just dependent upon ICT but also upon manual processes. Further, the same tools can be used to optimise the configuration of ICT to meet still higher level objectives, such as strategic scorecard objectives balancing cost with customer satisfaction. A commercial system of this type is a number of years away; however the technology is available.

9.8 Conclusions

Corporate organisations today are almost completely dependent on the availability and responsiveness of their networked applications. The interconnected and demanding environment in which they operate means that if a networked application slows down, so too does the organisation; if an application fails, work stops and the company loses money. Whilst historically the SLAs offered by telecoms providers have focused on individual network products, enterprise-scale customers are increasingly demanding SLAs for their critical business applications to provide predictable and consistent performance over the entire IT and networking infrastructure. Though the future promises sophisticated service-oriented architecture (SOA) frameworks for the negotiation and delivery of all application-level SLA requirements, other techniques must be applied shorter term to meet nearer business requirements. By optimising and orchestrating the ICT infrastructure to maintain the SLAs for key business applications, this chapter shows that a combination of state-of-the-art, yet current, technologies can deliver many of the requirements of an application-level SLA.

References

Azvine, B., Cui, Z, Majeed, B., and Spott, M. 2007 Operational risk management with real-time business intelligence, *BT Technology Journal*, Vol. 25, No. 1, January 2007, 154–167.

---

[4] www.nessi-europe.com.
[5] http://www.tmforum.com.

Carter, S.F. 2005 Quality-of-Service in BT's MPLS-VPN Platform, *BT Technology Journal*, Vol. 23, No. 2, April 2005.

Dames, M.P. Fisher, M.A., and Wittgreffe, J.P. 2005 Use of Policy Management to Achieve Flexibility in the Delivery of Network-centric ICT Solutions, *BT Technology Journal*, Vol. 23, No. 3, July 2005.

Dann, T., Gillam, J., and Thornhill, D. 2005 The Applications-Assured Infrastructure, *BT Technology Journal*, Vol. 23, No. 2, April 2005; also see the latest at bt.com under "large business and public sector" products.

Vitantonio, G.D., Legh-Smith, L., Miller, W., and Wilkinson, M. 'Meeting Business Objectives through Adaptive Information and Communication Technology'. *BT Technology Journal*, Vol. 24, No. 4, 2006.

Wittgreffe, J.P. and Dames, M. 2005 'From Desktop to Data Centre – Addressing the OSS Challenges in the Delivery of Network-Centric ICT Services'. *BT Technology Journal*, Vol. 23, No. 3, 65–78, 2005.

Wittgreffe, J.P., Trollope C., and Midwinter T. 2006 The Next Generation of Systems to Support Corporate Grade ICT Products and Solutions, *BT Technology Journal*, Vol. 24, No. 4, October 2006.

10

Mobility and ICT

Richard Dennis and Dave Wisely

10.1 Introduction

You do not have to be an ICT expert to acknowledge that almost all areas of our lives are becoming increasingly dependent on ICT in one form or another. Couple this with the impact of seamless, ubiquitous connectivity (fixed and wireless) on the growing human aspiration to live life wherever we are (mobility) and you have the backdrop for rapid change.

To establish the significance of mobile working, we consider trends in the UK public sector – a well-established ICT market. With an explosion in the number of mobile workers predicted and estimates that mobile workers already make up one-fifth (over a million) of all public sector employees, the pressure for ICT solutions to 'go mobile' is rapidly increasing. Annual investment on mobile working ICT solutions in the UK public sector is projected to rise to around £5 billion by 2015 as the market reaches a level of maturity. With an ageing population and the trend towards reduced hospitalisation requiring an army of mobile workers, health care is an obvious pressure point with mobile ICT services spend in the UK forecast by some sources to reach £1.5 billion by 2015 (Kabel, 2007).

10.1.1 Mobility – More Than Wireless Networks

In this chapter, we look beyond an approach where mobile simply equates to wireless and concentrate on mobility as an enabler. Sections 10.2 and 10.3 look at the problems and challenges that underpin the evolution of mobility. Acknowledging the importance of wireless technologies to mobility, Section 10.4 provides an overview followed by Sections 10.5–10.10, which attempt to expand how building out of connectivity can offer a platform for future mobile ICT solutions. The chapter concludes by extending the concept of mobility and mobile ICT solutions into the virtual space.

ICT Futures: Delivering Pervasive, Real-time and Secure Services
Edited by Paul Warren, John Davies and David Brown
© 2008 John Wiley & Sons, Ltd

Rapid mobile ICT growth without addressing the issues discussed in this chapter will result in increased complexity, reduced usability and a missed opportunity to engender the level of trust required for improved customer satisfaction.

10.2 Staircase to Success?

At a generic level, ICT could be viewed as simply providing answers to some fundamental questions for a technology based fulfilment process. It is all about 'knowing the customer' and attempting to resolve identity and relate to context . . . starting at the bottom . . .

How do we build an ongoing relationship?
What are they entitled to? . . .
What do they want? . . .
Who (what) and where are they? . . .

Technology has revolutionised the scale and scope of business by solving the problem of connectivity – voice/data communication. However, in doing so, it has unintentionally created another problem by removing some of the context required to make sense of it all! Why will a person in the high street asking for credit card details instantly arouse suspicion where a simple spoof email or web page appears to have much greater success?

Unable to use our natural ability, when communicating remotely, to resolve the wider context – identity, location, presence, etc – we are vulnerable. Without technology solutions to redress the balance, we run the risk of the requirements of control (security) suppressing a 'justified' level of trust. Establishing a justified level of trust (Cofta, 2007) based on an assessment of the context of a situation effectively kick-starts the interaction. Without establishing trust, society will be unable to maximise the benefits of being 'connected'.

10.3 Key Challenges

For successful ICT, accepting that mobility is part of the solution and the wider problem (the lack of context) encourages us to consider some of the key challenges before looking at the evolution of wireless mobility and its potential impact on future ICT solutions.

How can mobility related enablers support future ICT solutions?

- resolve the identity of the user(s) without compromising privacy;
  - *manage and facilitate multiple 'identities'*
- develop technologies that engender a justified level of trust;
  - *suppressing over and under trust*
- provide the additional context necessary to improve upon a face-to-face communication
  - *remote, online, virtual – technology enabled relationship.*

The balance between control and trust is currently tipped heavily in favour of control in the form of security solutions. Global reinforcement of the right to privacy will remain on the critical path to successfully redressing the balance. Designing trust into ICT solutions will offer significant differentiation in the market place.

10.4 Achieving the Foundations

Cellular mobile has been the main driver of mobile ICT – but things are changing with developments towards fourth generation mobile and the convergence of cellular and IP-based wireless technologies such as wireless local area network (WLAN) and worldwide interoperability for microwave access (WiMAX). A very brief history of mobile communications shows that this evolution has been going on for 30 years. Figure 10.1 shows a late 1970s scene with what might be called a zero generation mobile. Things have moved on – not only in fashion and hairdressing!

Table 10.1 outlines the mobile generations. Analogue cellular systems launched in 1980 lacked roaming, security and battery life.

Second generation (2G) digital systems were launched in 1990 with over 60% of all 2G handsets worldwide being GSM. 2G was about good quality mobile voice with roaming, secure encryption and improved talk and standby times. SMS – the now ubiquitous messaging service, happened almost by accident as it was originally intended as an internal communication system for GSM engineers. Even today, voice and SMS account for over 90% of the typical revenue of mobile operators in the UK (Van Veen, 2006).

Third generation networks were launched in Europe in 2001 promising users communication 'anytime, anyplace, anywhere'. The promise of high speed data, video

Figure 10.1   Zero-generation mobile

Table 10.1 Mobile generations

Systems	Products
High tier—cellular 1G—1980 launch	—advanced mobile phone system (AMPS)—USA —total access communication system (TACS)—Europe —analogue voice, insecure, no roaming
Cellular 2G—launched in Europe 1990	—IS-136 (TDMA) and IS95 (CDMA)—USA —GSM (TDMA)—Europe —personal digital cellular (PDC)—Japan —digital voice, roaming, low-rate data
Cellular 2.5G—1986	—general packet radio services (GRPS)—GSM enhancement—20–64 kbit/s —enhanced data rates for global GSM evolution (EDGE)—100 kbit/s —higher rate data
3G—2001—BT launch UMTS on Isle of Man (3G)	—mobile Internet —100 kbit/s

telephony, lower costs and a host of novel services as well as the 'mobile Internet' in reality has been slow to happen with voice and SMS still dominant.

Looking beyond Table 10.1, the key question for the mobile industry today is 'What comes next for cellular?' or 'What exactly is fourth generation (4G) mobile?' There is no industry or analyst agreement about the use of the term 4G. We believe convergence is the most likely path to 4G with recent announcements on WiMAX, WLANs (or WiFi) and terminal developments as well as key standard activities cited as evidence. In our view, convergence offers the best opportunities for mobile ICT services and synergies with fixed ICT. While the cost of basic connectivity will continue to fall, the key technologies, as well as the commercial battleground, will be in other areas of value-added services. Examples of potential value-add include identity management, quality of service (QoS), location-based services.

10.4.1 4G . . . 3G with a Turbo Charger?

3G was supposed to be about the 'mobile Internet', video calling and users watching football matches at bus stops, but ended up being about cheap voice and SMS (Lonergan, 2006). The cellular industry has taken a long time to get 3G handsets that users actually want: it is interesting to note that Apple's iPhone was not launched with 3G because of the power requirements and intellectual property licensing costs.

There have been a number of high-profile initiatives to increase revenues such as wireless application protocol WAP, picture messaging (multimedia messaging service, MMS), music downloads and instant messaging. These have had, at best, mixed results, with some success in the area of music downloads and disappointing revenues in most other areas. In an attempt to put things righ, 3G is being enhanced with new technology offering faster data rates and lower latencies – high-speed downlink packet access (HSDPA) in Europe and the Far East, and EVDO (evolution-data optimized) in the USA.

These 3.5G technologies offer practical downlink data rates of 1 to 2 Mbit/s with latency as low as 70 ms today. With cheaper data available to the user, the trend is away from a 'mobile special' Internet portal to partnering with existing players – Google, Microsoft and Yahoo (Mobile Gazette). The emphasis is on higher speeds, lower costs per Mbyte, revenue sharing agreements and corporate access.

Beyond HSDPA (2011–2015) is a new initiative called long-term evolution (LTE) with speeds up to 10 Mbit/s and an all IP network to reduce costs and improve integration with fixed networks and the Internet (Scrase, 2006). LTE is the cellular industry's response to WiMAX.

10.4.2 WiMAX as 4G

WiMAX is a technology generating a lot of interest – mostly because it is not a cellular industry initiative but a product of the computing community and standardized by the IEEE. Positioned as a 4G solution by some, WiMAX offers three key technologies that 3G and HSDPA do not have: an orthogonal frequency-division multiplexing air interface, all-IP network and the capability to use advanced antennas. Described as 'wireless local area network (LAN) on steroids' with large cells (up to 10 km), the standard (802.16e) offers features that are currently being 'retrofitted' to WLANs: QoS, strong security, voice over internet protocol (VoIP) support and handover between cells.

The key questions about WiMAX are commercial. When and by whom will it be introduced? What services will be offered? How will it compete with evolved 3G? With the existing cellular trend described earlier, WiMAX will have to offer higher bandwidths and lower costs to be successful – probably as low as typical digital subscriber line (DSL) prices to capture fixed Internet traffic.

Korea Telecom currently operates the world's only mobile WiMAX system (WiBro) providing VoIP, video calling, games, music downloads and Internet browsing. Whether this will tempt 3G and WLAN users remains to be seen. Key manufacturers are producing WiMAX integrated circuits and are bullish about including them in all future mobile computing devices.

10.4.3 4G Services

One approach to defining 4G is by means of a '4G service set'. Many people thought that 1G phones would only be used for emergencies – the voice quality would be so poor – that people would quickly find a landline to continue the conversation! The telecommunications industry's so-called 'killer application' was voice and for 2G, it was voice coupled with mobility. Users were prepared to pay a 'mobility premium' for a service they were already familiar with. The 3G service set describing the 'mobile Internet' turned out to be widely optimistic. Today, the trend is to define network capabilities that can be used by operator and third parties to develop a whole raft of innovative services. This is partly a realisation that, after years of fruitless searching, a 'killer application' for mobile data probably doesn't exist.

Figure 10.2 underlines the issues facing 4G services by plotting the revenue per Mbyte for services versus the required bandwidth – an approach used by Analysys (Heath *et al.*, 2006).

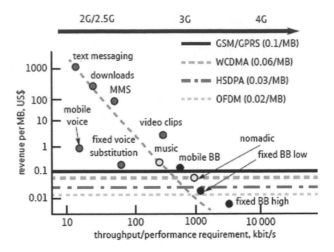

Figure 10.2 Revenue per Mbyte versus required network throughput (BB, broadband)

Most existing mobile services fall on a line of decreasing revenue per Mbyte with increasing network throughput. For example, users will pay 1.5× the cost of a voice call to add video but not the 10× the increase in capacity actually required, or a maximum £7 a month for mobile TV when the same volume of data charged as mobile voice would cost nearer £100 per month! The cost of supplying a Mbyte of traffic gets much lower on the newer networks (HSDPA, LTE and WiMAX), and the operator profit margin increases, although the potential reduction in operating range tends to offset the advantage. It is easy to see why services requiring little network capacity like ringtone download and SMS are so important to mobile operators and why they are keen to avoid cannibalization of these revenues.

10.4.4 Convergence as 4G

Fixed mobile convergence can be defined as seamless, ubiquitous connectivity and roaming. There has been a spate of convergence products – including BT's Fusion[1]. This offers seamless voice over WLAN and cellular access technologies for user (cost and indoor coverage) and operator (lower backhaul charges) advantage. Convergence currently means WLANs as the connection of choice because they are perceived as lower cost and are more compatible with the Internet. 3G exists for voice and when you are out of range of a WLAN.

For convergence to move beyond voice and SMS, WLAN access must become more ubiquitous and much lower cost than the current commercial charges. Although the business model remains unproven, there are signs of this happening with wireless cities springing up across the globe and many 'free' or open WiFi initiatives – such as that by FON with BT FON[2] potentially adding up to another one million UK consumer WLAN home hubs for public access.

---

[1] http://www.btfusionorder.bt.com/.
[2] http://www.btfon.com/.

10.4.5 Service Convergence

ICT users will increasingly demand more personalisation and seemingly more intelligent services if they are to be considered 'fit for purpose' replacements or improvements to existing solutions. Increased service intelligence builds on converged connectivity by enabling context aware services that are able to adapt to the underlying connectivity to ensure that the service is best delivered as well as best connected.

In principle, all-IP networks can offer more advanced services than those evolved around circuit-based solutions. Examples include smart call diversion – if your electronic diary indicates that you are in a meeting, your phone might automatically be set to vibrate for a select list of callers. If a client is calling about a specific product, the system might, for example, find the next available person with specialist knowledge of the product. With voice prices falling, solutions will offer this value-add at a premium allowing converged operators to differentiate and create revenue.

ICT services that can be started on one device, at one point in time, over any available connectivity and then resumed at another point in time using a different device/ connectivity – transparently and securely – are compelling when considering areas like health care or distance learning. Distance learning – a lifestyle choice in the developed world – is a given in developing countries like China. Here the barrier to western levels of further education is mainly the delay associated with the physical building programme required.

10.4.6 Lifestyle Convergence

Mobility will increasingly enable a 24 hr society with the resultant pressures on lifestyle and the provision of appropriate ICT solutions. A combination of mobility related enablers and increased service intelligence will lead to lifestyle services. The 'work–life' balance is an unmet user need and an example of the potential for lifestyle convergence tools and techniques. Convergence around the virtual enterprise, sales and field force automation and remote working are examples of 'mobile' ICT targets.

10.4.7 4G Wireless and Mobile ICT?

If the 4G landscape is going to be about multiple access technologies, then what are the key issues for creating and delivering ICT services over it? It will certainly not be about bit transport – mobile prices are falling relentlessly (in Denmark, for example, mobile telephony prices declined by around 40% within 1 year after discount providers had entered the market[3]), and WLANs are increasingly 'free' (Apple's iPhone has free access to 7500 access points in the UK run by The Cloud). In our view, the functions that sit between the content providers (including user-generated content) and the connectivity will be key – functions such as QoS, seamless connectivity and handover. However, the control of access to and the billing for services are prerequisite competencies for mobile ICT provision. If we are to move beyond web browsing and email in WLAN hotspots and beyond the stove-pipe cellular operator value chain, we need

---

[3] http://www.dbresearch.com/servlet/reweb2.ReWEB?rwkey=u21741525.

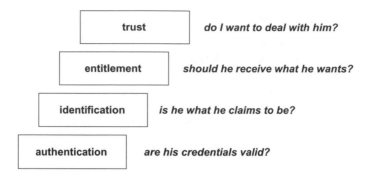

Figure 10.3   Building out of converged connectivity

to build out of the underlying connectivity. With established network concepts like QoS and handover well documented elsewhere (Wisely, forthcoming), we have chosen to explore this topic in more detail in the rest of the chapter.

10.5 Building Out of Connectivity

Figure 10.3 expands the staircase discussed in Section 10.2. Although ICT propositions will have specific requirements, the approach is generic and consistent with the common capabilities framework that underpins next generation networks like BT's 21CN[4].

10.6 Authentication – Do You Have the Key?

Discounting the locality of fixed network access, mobility requires alternatives and it finds them in an array of service-oriented authentication mechanisms.

With around six billion SIM (subscriber identity module) cards worldwide (SIM Alliance, 2007) SIM based authentication is a proven technique that can be expanded still further to provide secure access in a seamless converged world. Previously wired or fixed devices will increasingly become wireless and capable of performing the authentication to fixed and mobile networks. Apart from SIM based authentication, software SIM, trusted computing platforms (Dinesh, 2005) and strategies using smart-cards and Smart Media Cards (Helixion, 2008), all have strengths in certain applica-tions and markets.

Currently, the majority of authentication is limited to unlocking connectivity. The ultimate goal is to validate access to the higher layers – the network intelligence/IMS (3GPP, 2007). Higher layer or service based authentication allows differentiation by added features such as service based QoS, which represent higher value to customers. Exceptions exist – machine to machine services for example – where connectivity may be the only mechanism for revenue generation.

With the familiar personal identification number (PIN)-based mobile authentica-tion, the device is authenticated – not the individual using the device. Authentication

---

[4] http://www.btplc.com/21CN.

of the individual implies biometrics and, although an emerging but problem area, they remain valid components of the authentication hierarchy. The specific level of authentication required depends on the risk associated with a successful authentication.

10.7 Identification – Who or What Are You?

Establishing identity in ICT solutions today tends to be on a service by service basis with no attempt to make it seamless or sticky. Having accepted that the user needs to be authenticated, an identity will have been established, although it may not need to be fully resolved as explained later.

Establishing a network-based identity has the added advantage of being able to share a single authentication mechanism with all of the devices used by, or associated with, that identity. The personal communication environment (PCE) (Fallis, 2007) uses proximity to expand the concept and acknowledge that more than one device could be used by a single identity simultaneously. In an office situation, for example, any monitor could automatically display the contents of your handheld device simply because it becomes part of your PCE. As a shared resource, the monitor may only require a valid corporate identity to allow automatic access. Membership of the corporate identity is sufficient with no requirement to resolve the individual's absolute identity.

When consuming content, the associated rights can be assigned to an individual identity rather than to a specific device allowing it to be moved and consumed within the restrictions of the original agreement. With the growing trend for user-generated content (e.g. YouTube, 2007), distribution becomes increasingly based around a group identity. A group (friends, employees) interested in or authorised to consume local news, events, gossip, traffic problems or enterprise news, updates, and so on.

10.7.1 Multiple Identities

In an online society, an individual can have any number of alternative identities. This rapid increase in the number of multiple identities reflects the ease with which they can be adopted and disregarded as much as the need for privacy or anonymity. The undesirable outcome is the endless raft of naming, addressing, numbering and passwords associated with modern living.

Finding solutions to minimise the risk without increasing the control or security to a point where the offering becomes unusable is an important challenge. Accepting that people will continue to adopt a vast array of different identities to maintain their privacy and anonymity is essential. A payment identity (bank account) will have been completely resolved at the application phase. However, increasing identification fraud means the trend towards two-stage authentication is likely to replace a static payment identity.

10.8 Entitlement – What Can I Allow?

Entitlement is the gatekeeper between request and fulfilment and is the cornerstone for the majority of ICT solutions. Currently, a shared or group identity is often the determining factor when confirming entitlement. Increased context should mean the risk associated with a request from an individual can be better quantified to make a more 'considered' decision. In some ICT solutions, entitlement does not necessarily resolve to an individual identity or to an identity at all. Ticketing, for example, establishes entitlement without linking back the individual identity, with the notable exception of flight tickets and high-profile events where the individual identity is preserved.

In some existing applications like affinity cards, establishing identity is a double-edge sword. The system often holds only basic knowledge about an individual for fear of violating civil liberties and regulation. An additional entitlement lure (such as an extra 10% discount) provides the opportunity for further differentiation by collecting much more detailed information while maintaining the illusion of the enhanced value of the relationship.

10.9 Trust – How Do We Maintain This Relationship?

Trust has numerous definitions (Cofta, 2007), but it is a concept that humans have evolved to value.

Trust is arguably the most important factor when building a relationship and is the ultimate differentiator for a service company. A competitor may be successful in duplicating the processes, procedures, services and brand values, but it can never duplicate a trusted relationship. It can attempt to build a new relationship – but there is normally considerable resistance unless the existing relationship has already failed. Organisations remain very protective about the data mining potential of their relationship with their customers for fear of a consumer-led backlash. Once again, privacy and trust are key to the user's perception of a worthwhile relationship.

One approach is to attempt to mirror in the virtual world something of the natural, inbuilt, human approach and understanding of trust in everyday life (Lacohée et al., 2006). Research suggests that the value of a trusted business relationship can be quantified and if you can measure it – you can improve it.

The goal is to unlock the value of a trusted relationship using the technology associated with authentication. The more contextual information available, the easier it is to determine the potential risk of meeting the request. Even relatively small steps in this direction may provide an opportunity to redress the balance between control and trust (Cofta, 2007) and to explore alternative business models.

10.10 Design for Trust

Research into trust enhancing technologies (TET) attempt to build additional context so that the risk can be assessed and acted upon (Cofta, 2007). In future 'design for trust', ICT solutions will be able to apportion charges by the cost of servicing the risk encountered.

eBay could be considered as a simple example of a TET actually working today (Cofta, 2006). Without the quantification and presentation of trust based information on validated identities (seller/buyer), it would be difficult to see how the eBay model could operate – let alone be successful. Fraud exists, but it is considered insignificant and largely ignored. eBay is considered an easy, fun way to offer unwanted items to a large audience of potential buyers. It is difficult to see how increased control would make the service more appealing: increased trust, resulting in reduced risk based on the approach discussed in this chapter, could give a wider appeal and increase the range of applicable markets.

Automatic and largely invisible, the TET of an ICT solution is likely to be more consistent with the 'Intel inside' sticker than the eBay service itself. In the same way that the sticker says something about the quality and potential of the experience that awaits the PC user, use of 'TET' could indicate a model consistent with the increased confidence of a trusted relationship established between the user and the ICT service.

10.11 Virtual Worlds and Second Earth

So far, we have concentrated on the physical world – the obvious focus for today's mobile ICT solutions. True mobility means the distinction between the physical and virtual world becomes increasingly blurred with mainstream acceptance fuelled by a highly mobile, always connected generation. The opportunities and requirements for businesses and governments to follow its customers into the virtual world are compelling.

The projected emergence of phenomenon like Second Earth (Roush, 2007) and its need for ICT is worth exploring. Second Earth is in essence the convergence of Second Life (Second Life, 2007) and Google Earth. It is a technology enabled world we could all inhabit that will be at best a virtual representation of planet Earth and at worst, the frighteningly deep and dark spaces of collective human consciousness without the shackles of convention surrounding an evolved society!

It is easy to dismiss the concept as pure escapism or a youth based gaming paradigm, but just like the real earth, the opportunity for human survival, interaction and enrichment means there is scope for society, commerce, etc., and significant riches for those who successfully tame this wild new frontier.

We believe that the concepts of mobility, identity and trust described in this chapter are just as applicable in a virtual world as they are to someone conducting eCommerce today. The principle is one of using technology to provide or enhance context in an environment where it is weak or no longer exists for one reason or another. Although anonymity may be an obvious attraction to many of the early adopters, Second Earth society is likely to rapidly evolve by benefiting from identity and trust built on connectivity – effectively the bedrock and life blood of a virtual world. Until Second Earth is a reality, Second Life provides an interesting and challenging testbed for research in the area of trust based models.

Given the desire, it is not difficult to see how a digital nervous system of sensors could translate the physical world into a virtual equivalent to form an augmented reality. However, simple translation from the physical to virtual fails to grasp the potential for new 'mobile' ICT solutions based on data and information representations only possible in a virtual world.

Major corporations and organisations hot on the trail of the next Internet have already shown considerable interest in Second Life. Second Earth offers even more potential, sweeping away the well-established channels to market and potentially generating new business models.

10.12 Heaven or Hell?

Starting with the relationship between mobility and ICT, the authors have explored how enablers like identity and trust could build out from connectivity to facilitate improved and more mobile ICT solutions.

Much of the underlying technology required to facilitate a 'designed for trust' approach already exists.

The challenge will be to engineer solutions that can generate differentiation and the required profit while encouraging trust, privacy, inclusion and the re-establishment of community. The dangers exist that the privileged will be trusted and enablers like identity will simply increase the gap between people that have and those that do not. Surveillance of the population is increasing all the time, with high resolution satellite cameras and a rapidly growing numbers of fixed surveillance cameras capturing many UK citizens up to 300 times per day. Couple this with increasingly sophisticated video processing and facial recognition software, and maybe the only difference between you and a Big Brother (Channel 4, 2007) contestant is editing and entertainment value.

Increasingly, the link between mobility and identity will tend to become much more about the searching, sorting, manipulating and delivering 'information' over a seamless hierarchy of converged connectivity. If we get it right, we can look forward to an efficient and effective raft of ICT based services resulting in a stronger, trusted relationship between customer and supplier. If not, then the trademark 'Google Earth' may become the ultimate irony: Google's search engine legacy could evolve along the lines discussed in this chapter to become a dominant force in an information rich society, powerful enough even to play a role in the political determination of the future of planet Earth – first and second . . .

References

3GPP, 2007. Internet Multimedia Subsystem (IMS), http://www.3gpp.org/ftp/specs/html-info/23-series.htm, visited on 12/11/07.
Channel 4, 2007. Official Big Brother UK Website, http://www.channel4.com/bigbrother, visited on 09/11/07.
Cofta, P, 2006. Confidence Creation Framework of eBay. Paper presented at Networking and Electronic Commerce Research Conf NAEC2006 October 2006.
Cofta, P, 2007. Trust, Complexity and Control, John Wiley & Sons, Ltd. ISBN 9780470061305.
Dinesh, K, 2005. Trust in Trusted Computing – the End of Security as We Know It, Computer Fraud & Security December 2005.
Fallis, S, 2007. Pervasive Information: the Key to 'True' Mobility, BT Technology Journal, Mobility and Convergence, 25 (2) 179–188.
Heath, M, Brydon, A, Pow, R, Colucci, M and Davies, G, 2006. 'Prospects for the Evolution of 3G and 4G', Analysys.

Helixion Limited, 2008. http://www.helixion.com/, visited on 13/02/08.

Kabel, 2007. Flexible and Mobile Working in the UK public sector, http://www.kable.co.uk/kabledirect/.

Lacohée, H, Crane, S and Phippen, A, 2006. Trustguide, http://www.trustguide.org.uk/Trustguide%20-%20Final%20Report.pdf.

Lonergan, D, 2006. 'Driving usage of value-added mobile services', Yankee Group.

Roush, W, 2007. Second Earth Technology Review, MIT, July/August.

Scrase, A, 2006. '3G Long Term Evolution Overview of the Current Status of 3G LTE Standards', CTO, ETSI Informa 3G LTE Conference, The Café Royal, London, UK (October 2006).

Second Life, 2007. Linden Research, Inc., http://secondlife.com/

SIM Alliance, 2007. http://www.simalliance.org/, visited on 12/11/07.

YouTube, 2007. http://www.youtube.com/, visited on 12/11/07.

Van Veen, N, 2006. 'Getting Consumers to Use Mobile Services', Forrester.

Wisely, DR, 2008. *IP for 4G*, John Wiley & Sons, Ltd. ISBN-10: 0470510161.

11

Pervasive Computing

David Heatley, George Bilchev and Richard Tateson

11.1 Pervasive Information

Look around today and you will see numerous examples of pervasive information. These range from the popular pastime of 'Googling on the net', which is highly informal and (in principle) open to everyone, through to dedicated search engines and databases with password protected access that support major organisations in specialist closed applications. A good example of the latter is the Care Records System[1,2] which is being progressively rolled out across the National Health Service (NHS) in England and, when fully in service around 2010, will be a key element of all electronic patient information handling across the country in the public sector. Clinicians can access information on patients from (in principle) any computer terminal connected to the National NHS Network[3] via their local intranets, for example, in a hospital, community clinic or doctor's surgery. The content of that information is drawn from a variety of dispersed sources which are brought together in a single virtualized form to deliver a richness that has been hitherto impossible. For example, the information set for any particular patient could comprise a mixture of text and detailed graphical data (e.g. radiology images, ultrasound scans, etc.) derived from many consultations over time by different clinicians and from different health centres, all of which can be accessed (by approved users) in real time from anywhere within the NHS.

Today's pervasive information systems such as the example above are relatively prescriptive in that they operate in a particular way, and the information they access is captured and delivered in a particular way, all of which users must adapt to. It seems unlikely that any one, or a combination of these pervasive information systems, will

---

[1] http://www.connectingforhealth.nhs.uk/.
[2] http://www.bma.org.uk/ap.nsf/Content/ncrsguidance.
[3] http://www.n3.nhs.uk/.

ICT Futures: Delivering Pervasive, Real-time and Secure Services
Edited by Paul Warren, John Davies and David Brown
© 2008 John Wiley & Sons, Ltd

evolve to the point where it becomes globally standardized in terms of architecture, information handling, etc. However, the full potential of pervasive information will not be realized unless there is harmonization of technology from a vendor and service provider perspective, and interface from a user perspective. In order to achieve some degree of harmonization across a broad range of (or even all) pervasive information systems, today's architectures need to be replaced by more open and flexible architectures which allow information to be composed from a multitude of repositories as required. Research is underway on developing a pervasive information architecture (PIA) (Fallis *et al.*,2007) which aims to be agnostic to the kind of information required by individual users, the way in which that information is sourced and compiled, and how it is delivered. The PIA and similar emerging schemes are also key enablers for truly mobile working, where users can (in principle) access any kind of information, via any kind of portal, from any information source, at any time, from anywhere, etc. The demand for that degree of mobility already exists, particularly in the public and service sectors, and that will help to drive its development.

Of course, with pervasive information and mobility, where people can access information from anywhere on any device, etc., comes the need for identity authentication, particularly if the information is necessarily restricted to approved users. Today's solutions, such as the smartcard and personal identification number (PIN) system used across the NHS,[4] although effective, do not generally support the degree of mobility alluded to here, and hence represent a barrier to the full potential of pervasive information being realized. Novel alternative schemes are beginning to emerge in which the client's authentication token is integrated into a portable device such as a personal digital assistant (PDA), and these promise to be eminently more suited to accessing pervasive information in a truly mobile environment.

The primary research challenges going forward for pervasive information are therefore to develop new architecture concepts such as PIA as well as the new technologies described in the rest of this chapter, and to create an 'information environment' that embraces everyone and delivers what they want, securely.

11.2 Pervasive Sensing

11.2.1 Overview

Pervasive sensing is the ability to acquire information about almost every object, person or environment. It is fueled by the proliferation of small electronic devices embedded in everyday objects, devices which range from radio frequency identification (RFID) tags on items purchased in shops to environmental monitoring sensors and control actuators in homes and offices. Pervasive sensing also extends to small ICT devices carried by people which have innate sensing capabilities, such as mobile phones and PDAs which can sense location and temperature, and sometimes incorporate accelerators to monitor movement.

Pervasive sensing has the potential to link the digital world of computing and the Internet to the physical world of sensory data such as temperature, pressure or acceleration, and so create new value from hitherto unavailable information sets. It is the

---

[4] http://www.connectingforhealth.nhs.uk/systemsandservices/rasmartcards.

main information source for creating pervasive computing contexts and context aware applications.

Pervasive sensing is a natural evolution from the area of sensor networks, where specific features of interest are observed, alarms raised and appropriate actions taken. An example of a sensor network application in the health sector lies in telecare for the management of chronic diseases (Reeves *et al.*, 2006), where people are monitored both in the way they behave (e.g. using sensors embedded into people's homes) and their physiological status (e.g. using body worn sensors and body area networks). Alarms are automatically raised based on observed activities of interest or deviations from the norm, which signify a possible deterioration in a person's well-being. In the current generation of sensor networks, the functionality of the sensors and actuators, the IT infrastructure and the business logic are all specific to a particular application, such as telecare. Consequently, a home security system cannot share the motion sensors from the telecare system, and so on. This has led to a proliferation of bespoke closed systems.

Pervasive sensing going forward aims to create open sensor networks, where information from sensors can be shared and re-used by multiple applications. In this respect, pervasive sensing is not dissimilar to the concept of the sensor web (Botts *et al.*,2006). This concept was first voiced during the 1990s: millions of connected online sensors monitoring the physical world, the sensor capabilities being described using metadata so they can be published and understood by anyone with web access and appropriate authentication. This model is similar in concept to the WWW where a web browser can access this vast information space thanks to the adoption of key standards such as HTTP, HTML and XML. In 2001, a data modelling language for sensors, SensorML, was introduced into the Open Geospatial Consortium[5] which eventually spawned a new working group called Sensor Web Enablement (SWE).[6] The group was tasked to produce a framework of open standards for web connected sensors and all types of sensor systems. Fast forwarding to today, SWE has now released an open set of 'Issue 1' standards which have been through many drafts and practical evaluations using prototype deployments.

The more ambitious aims of pervasive sensing are to:

- capture the opportunity that low cost sensors present by monitoring, e.g. the environment or the industrial infrastructure in real time, and deliver significant monetary value through cost savings in, e.g. reducing power theft, increasing crop yield, etc.;
- stimulate a commercially viable market for intelligently handling sensor data from multiple sources and fusing (combining) it in innovative ways to create and deliver novel advanced service offerings and applications;
- promote an open innovation environment for diverse innovative applications and solutions, which is based on best practice (including emerging standards) and encourages wide participation from the public by publishing sensors via the web and sensor mash-ups;
- raise awareness of the potential impact of emerging pervasive sensing technology on the economy, industry and lifestyle of people among policy and decision makers in government and industry.

---

[5] http://sensorwebs.jpl.nasa.gov/.
[6] http://www.opengeospatial.org/projects/groups/sensorweb.

11.2.2 Challenges in Pervasive Sensing

A key challenge is the development of an overall architecture that aspires to ultimately support the deployment of large-scale pervasive sensor environments. This involves understanding and solving the fundamental design, deployment and operational challenges for large-scale pervasive sensing networks, and providing the necessary tools to address the issues. A key goal will be to support sustainability and re-use of such large-scale pervasive sensor environment infrastructures. This architecture will be required to

- provide a pervasive sensor environment that is utilised by many infrastructural and value-added applications and service providers;
- accommodate multiple data formats through the virtualisation of millions of pervasive sensors;
- support the sharing of sensor information between multiple domains over a common sensor infrastructure using open interfaces based on best practice and/or emerging standards;
- provide an 'open' and extensible sensor environment and a 'public' interface to allow anyone to add sensors to the platform;
- deploy security and privacy solutions where appropriate such as virtual private network (VPN)-like secure overlays and applications;
- have the potential to support a wide range of applications, sectors and end users via a toolkit.

11.2.3 Pervasive Sensing Infrastructure

In order to create an end-to-end IT architecture for pervasive sensing, the following six key areas need to be addressed.

11.2.3.1 Physical Environment

The physical environment is where the devices are embedded and deployed, and where there is a great deal of diversity in functional requirements. This stems directly from the diversity of the physical environment itself: buildings, tunnels, bridges, human body, cattle, fields, the sea, glaciers, etc. The sensors, devices, their communication protocols and execution environments have ranged from custom purpose-built hardware and communications to using one of many proposed embedded platforms (e.g. Crossbow motes[7]) and communications protocol suites (e.g. Zigbee[8]). In order to develop a generic solution, attempts are being made to transfer application platforms from mainstream IT to the sensor devices. For example, all major platforms have a small footprint version that could run on sensors/motes. Intel's motes[9] run on a small

---

[7] http://www.xbow.com/Home/HomePage.aspx.
[8] http://www.zigbee.org/en/index.asp.
[9] http://www.intel.com/research/exploratory/motes.htm.

footprint Linux; Sun uses the Java platform for their SunSpots,[10] and Microsoft has a .Net micro-version[11] that has been ported to Crossbow motes.

11.2.3.2 Gateway Environment

The gateway environment acts as the bridge between the IT infrastructure and the specific pervasive sensing deployment environment. It links the internet protocol world with the local connectivity technology (i.e. Zigbee, IEEE 802.15.4, etc.) and presents an opportunity to create a virtualisation layer abstracting the specific hardware, software/firmware and local communications of the pervasive sensing edge devices. It is assumed that the gateway environment is computationally more capable than the edge devices and can run the virtualisation layer.

Virtualisation could be achieved by hiding the specifics of the deployed systems behind a set of standardized application programming interfaces (APIs). For example, if the deployed system is for environmental monitoring, open sensor standards (such as the standards developed by the SWE initiative mentioned earlier) could be used to describe the interfaces to read sensor data. Such a virtualized open sensor layer would also be semantically more descriptive by utilising well-established schema such as for observations and measurements,[12] thus enabling more dynamic applications that could discover resources and generate sensing plans on-the-fly. A typical example of such applications would be disaster management/recovery where resources (i.e. sensors) are dynamically called upon based on the specific situation.

11.2.3.3 Information Transport

The information transport environment is the medium through which data pass from the edge devices/sensors (i.e. the source) to applications that use the data (i.e. the sink). It takes care of routing, scalability, resilience, quality of service (QoS) (i.e. guaranteed delivery), etc. There are various levels at which information transport can be described. The lowest meaningful level is the packet level where routers are aware of the message content and route accordingly. The next level up is the logical stream level, where streams of packets form topic queues and applications subscribe to them. This forms an overlay network tasked to provide the necessary routing, scalability and QoS, perhaps with the help of a broker network. Some systems claim to operate without a broker[13] (i.e. a peer-to-peer basis); however, they assume the existence of reliable multicast to announce the existence of peers and the topics they handle, which may not always be the case in a real setting.

11.2.3.4 Information Processing

The information processing environment is responsible for translating raw sensor data into useful information that can be acted upon, such as events and alerts. This could

---

[10] http://research.sun.com/spotlight/SunSPOTSJune30.pdf.
[11] http://msdn2.microsoft.com/en-gb/embedded/bb278106.aspx.
[12] http://www.opengeospatial.org/standards/requests/37.
[13] http://sourceforge.net/projects/mantaray/.

be achieved by applying various information processing actions such as fusion, filtering, correlation, etc. Information processing could be done in a distributed fashion in conjunction with the information transport, i.e. as information streams on the network travel through various points such as brokering or mediation nodes, they could be processed. Generic complex event processing engines could be used to provide a generic information processing environment.[14,15]

11.2.3.5 Enterprise Integration

In this area, data about the physical world from sensors/edge devices are integrated with the operational requirements of the enterprise. Real-time business intelligence (RTBI) investigates how the data about the physical world support and add value to the business processes, and also add an analytics layer. Vendors such as Oracle and BEA[16] are now starting to offer complete technology stacks aimed at enterprise sensing applications which include RTBI, business analysis and business monitoring.

11.2.3.6 Application Environment

The application environment is a helper tool to provide applications with additional functionality. Consider, for example, a discovery mechanism for locating objects or sensors (and their capabilities) or a geospatial reasoning engine allowing queries to include spatial references. The application environment could offer these and other capabilities such as a visualisation of the multiplexed physical and virtual spaces and a whole set of communications and convergence services such as BT's Web 2.0 software development kit.

11.3 Pervasive Intelligence

11.3.1 The Many Facets of Pervasive Intelligence

Pervasive intelligence brings together elements of pervasive computing whereby a disperse network of many autonomous or semi-autonomous devices carry out the tasks of data gathering, rule execution (sometimes called 'reasoning') and action. This generic description embraces a very wide range of specific instances. The data gathered, for example, could be temperature readings from a thermometer or a mouse click in a user interface. The 'disperse' nature of the network could be geographical or based on a network overlay.

Pervasive intelligence is most likely to be more effective than alternative control strategies when:

- information is most usefully gathered at the edges of the network, either via sensors or other data input (e.g. user input);

---

[14] http://www.coral8.com/.
[15] http://www.streambase.com/.
[16] http://www.bed.com/.

- communication with a centralised decision maker is problematic (perhaps because it is slow, unreliable or expensive, or where there are real or perceived privacy and security issues);
- action can be taken at the edges;
- actions taken at the edges have been designed to contribute to overall system success (the action could still seem 'selfish' to the autonomous nodes).

The effects of pervasive intelligence vary widely in how overtly they can be perceived by human users of the networks in question. In some cases, pervasive intelligence achieves deeply hidden network management functions, which are only evident to human users, if at all, through the performance of the network. In other cases, the 'actions' taken by the pervasive elements may be presented directly to human users, perhaps by changing the content displayed on video screens in a public place.

11.3.2 Sensors Sharing Information to Coordinate Environments for Humans

Pervasive intelligence is contributing to the next generation of building automation. In its current form, building automation is the control of air conditioning and lighting in buildings through a programmed network of electronic devices. It is motivated primarily by a desire to minimise energy costs by ensuring that heat (or cooling) and light are provided only when and where needed.

Current research work is pushing this concept further in the pervasive direction by incorporating a much greater variety of sensors to gather more detailed information about the activities of people in the building, and by aspiring to use the information not merely to minimise energy consumption, but to maximise the well-being and effectiveness of the people therein.

One example of this is the recent Building Awareness for Enhanced Workplace Performance (Bop!) project,[17] an academic/industrial collaboration supported by the Technology Strategy Board of the UK government. BOP aimed to combine information from sensors monitoring the activity of the building occupants with information proactively supplied by the occupants themselves through an interface (where those interfaces are designed to be 'playful' and nonintrusive). These data gathered with and without explicit interaction provide information about perceptions, moods and working styles of the people in the building.

Data gathering in the BOP style could be widely used in places that would benefit from a better understanding of the quality and dynamics of a space – for example, in supermarkets to monitor customer flow, in work places or office buildings to quickly locate available/occupied rooms, at petrol stations as a means for customers to provide rapid feedback to service suppliers (for example, on the state of public toilets), or even in prisons to give early warning of unrest.

Within the BOP project, the information gathered is delivered to people who are engaged in monitoring, modifying and designing buildings. The next step in the pervasive intelligence story for this area is to complete the automatic feedback loop and to use the information to directly control aspects of the building environment which

---

[17] http://www.ieeexplore.ieee.org/ocpl/freeabs_all.jsp?arnumber=4126020.

might change the perceptions, moods and working styles of the people being monitored.

11.3.3 Wireless Devices Gossiping by Radio to Maintain and Adapt the Radio Network

In this example of pervasive intelligence, the system is being manipulated (or is manipulating itself) at a low level in the communications architecture. The radio channels supporting communication among the various mobile devices in an ad hoc network are themselves being chosen by the mobile devices according to relatively simple rules applied to information gathered locally by the device.

To operate a radio network for cellular mobile communications, the limited resource of radio channels (whether defined by frequency, timeslot or code) must be allocated to all the nodes of the network (Krämling et al., 2000). A good allocation will provide nodes with adequate resources to meet local demand while also keeping interference between adjacent channels below acceptable levels (Lee, 1989).

Much work on this computationally challenging problem of channel allocation has focused on planning for an unchanging topology and pattern of demand using empirical or simulated interference data. However, as networks become larger, the case for pervasive intelligence (in the form of decentralised local control) becomes more attractive (Akaiwa and Andoh, 1993). For mobile ad hoc networking where network topology changes rapidly and unpredictably over time, some form of decentralised self-organising control is essential (Charter, 1999; Holliday, 2002; Hubaux et al., 2001). The pervasive intelligence approach to this problem is an example of 'online' dynamic optimisation in which channel use is reallocated in response to changes in demand and interference as they occur.

One example of such a method (Tateson, 1998) is inspired by the local negotiations among the cells of a developing fruitfly which lead to the successful production of a precise global pattern in the resulting adult fly. By reacting to inhibitory signals from neighbouring nodes, a node modifies its choice of channel. When all nodes in the network follow the same algorithm, good channel allocation plans can be produced without any need for centralised planning or measures of global optimality. This method has been applied in simulation to a mobile phone network, which is one example of the general class of wireless communications networks. The method has also been extended (Tateson et al., 2003) to address dynamic planning in a military communications network where the unpredictable nature of the network makes dynamic planning especially appropriate.

Depending on the details of the radio network, the fruitfly-inspired method may not adequately control the radio network, essentially because the data gathered (on interference from neighbouring devices) leave each device potentially 'blind' to some significant sources of interference. For these situations, a modified control system, also meeting the definition of pervasive intelligence, has been proposed. Dynamic interference nullification (Saffre et al., 2004) uses randomised frequency/time-slice selection, coupled with variable transmission energy levels and adaptive preference for interference-free channels in order to 'evolve' a network of reliable wireless links. The objective is to converge to a state where the rate of collisions at the receiver's end is kept as low as possible, so as to ensure that all nodes in the network are capable of exchanging information over a 'clean' data-link.

11.3.4 Information Shared to Ensure Services are Available to Users

For a very different example of pervasive intelligence, consider the peer-to-peer provision of services. Service-oriented computing (Muschamp, 2004) is now a well-established mechanism for offering computing services over networks. The standard current approach is to rely on a central brokering service which receives requests from 'user' nodes and supplies the address of a 'provider' node able to supply a service to meet that request. In some situations, the use of a centralised brokering service can be problematic. For example, if a third party wishes to disrupt the service (which could be the case in a military battlefield scenario, or even in a commercial scenario akin to the peer-to-peer file sharing of Napster), compromising the centralised brokering service will destroy the function of the whole network. This vulnerability creates a niche for a pervasive approach whereby the discovery of services is mediated by the peers rather than by a dedicated broker, and hence there is no single point of failure to be targeted.

One example of current work in the area of pervasive intelligence for service-oriented computing is the NEXUS middleware which forms part of the Hyperion project (Ghanea-Hercock *et al.*, 2007) prototyping next-generation battlespace information services. NEXUS seeks ways of fusing the service-oriented and pervasive computing paradigms in order to build intelligent, robust and resilient networks connecting dynamic islands of service resources.

The services provided by NEXUS can be very diverse, but for clarity, it is easiest to focus on intelligence services available on the battlefield. So for example, a commander may desire a video feed from an unmanned aerial vehicle showing real-time information on a specific part of the battlefield. In a traditional service-oriented approach, the commander would send a request for such a service to the central broker. Provided such a service was registered with the broker, the broker could then provide the address of the network device hosting that service. In the NEXUS approach, the commander queries 'peers' – other nodes in the network which are themselves acting as requesters and/or providers – and the service provider advertises its existence also via its peers rather than a broker.

In the event that the video feed in question fails (perhaps because the vehicle is destroyed, taking the service offline, or perhaps because the vehicle has flown on to a different part of the battlefield, meaning that its service description no longer matches the request), the commander will be offered a list of substitute services. The list is ranked according to the closeness of fit between the service and the requirements of the commander, so for example, a recent text intelligence report on the correct area of the battlefield could be more useful than a still image from a satellite.

The pervasive approach brings challenges as well as avoids weaknesses. It is now up to each service to advertise its existence in an accurate way (what service is it offering?) and in a timely way (is it online and functioning at the current time?). Methods to support service description and real-time monitoring are essential parts of the continuing research work.

11.4 Closing Remarks

This chapter has attempted to show, firstly, that pervasive computing is already integral to society in many key areas, and the pervasiveness of the technology and the magnitude of its impact is certain to grow, and secondly, the great diversity with which pervasive computing is being applied, from patient monitoring and information gathering in health care applications through to battlefield scenarios. Several challenges exist going forward. The challenge for researchers in the field is to develop technologies and functionalities that are truly scalable and agnostic to all operational circumstances and user requirements. The challenge to network operators and service providers is to deploy these technologies and to exploit their capabilities in ways that deliver a seamless pervasive experience to users. And the challenge to users is to use the continually increasing access to pervasive information and services to enrich their lives and to support society in its wider goals.

References

Akaiwa, Y. and Andoh, H. 1993, Channel Segregation-A Self-Organized Dynamic Channel Allocation Method: Application To TDMA/FDMA Microcellular System. *IEEE Journal On Selected Areas In Communications* 11(6), 949–954.

Botts, M., Percivall, G., Reed, C. and Davidson, J. 2006, OGC Sensor Web Enablement: Overview And High Level Architecture. Open Geospatial Consortium Inc. (OGC) 06-050r2 White Paper (July 2006).

Charter, W. G. 1999, *Mobile Ad Hoc Networks (MANET)*, IETF, 1999.

Fallis, S., Payne, R., Limb, R. and Allison, D. 2007, Pervasive Information – the Key to True Mobility. *BT Technology Journal* 25(2), 179–188.

Ghanea-Hercock, R., Gelenbe, R., Jennings, N., Smith, O., Allsopp, D., Healing, A., Duman, H., Sparks, S., Karunatillake, N. and Vytelingum, P. 2007, Hyperion—Next-Generation Battlespace Information Services. *The Computer Journal* 50, 632–645.

Holliday P. 2002, *Self Organisational Properties of Military Ad Hoc Wireless Networks*, Land Warfare Conference, 2002.

Hubaux J. P. *et al.* 2001, Toward Self Organized Mobile Ad Hoc Networks: The Terminodes Project. *IEEE Communications* 39(1), 118–124.

Krämling, A., Scheibenbogen, M. and Walke, B. H. 2000, Dynamic Channel Allocation in Wireless ATM Networks. *Wireless Networks* 6, 381–389.

Lee, W. C. Y. 1989, *Mobile Cellular Telecommunications Systems*. McGraw-Hill Book Company, New York.

Muschamp, P. 2004, An Introduction to Web Services. *BT Technology Journal* 22(1), 9–18.

Reeves, A. A., Ng, J. W. P., Brown, S. J. and Barnes, N. M. 2006, 'Remotely Supporting Care Provision for Older Adults' in Wearable and Implantable Body Sensor Networks International Workshop, BSN 2006, MIT U.S. (April 2006).

Saffre, F., Tateson, R. and Ghanea-Hercock, R. 2004, Reliable Sensor Networks Using Decentralised Channel Selection Source Computer Networks. *The International Journal of Computer and Telecommunications Networking* 46(5), Special Issue: Military Communications Systems and Technologies, 651–663.

Tateson, R. 1998, *Self-organising Pattern Formation: Fruit Flies and Cell Phones* in A. E. Eiben, T. Back, M. Schoenauer and H-P. Schwefel (eds.) *Proc. 5th Int. Conf. Parallel Problem Solving from Nature*, Springer, Berlin, pp. 732–741.

Tateson, R., Howard, S. and Bradbeer, R. 2003, *Nature-inspired Self-organisation in Wireless Communications Networks*. Post-proceedings of ESOA 2003.

12

Artificial Intelligence Comes of Age

Simon Thompson

12.1 Introduction

Artificial intelligence (AI) is a very old idea; for example, the myth of Pygmalion reflects the fascination that creating an intelligent being had even for ancient peoples. The invention of digital computers in the middle of the 20th century provided philosophers and mathematicians with a tool to model the processes of intelligence and a way to test their hypothesis about the mechanisms of reasoning that underpin human logics, allowing scientific method to be applied. John McCarthy of Stanford University coined the term *artificial intelligence* in 1956 in response to the failure of earlier attempts to organize workshops on the topic using other titles (Lighthill 1973).

Fast progress was made. Chess playing was at the time seen to be one of the pinnacles of intellectual achievement, and so, early researchers adopted the development of a competent computer chess player as a key challenge problem. Against all predictions, a credible artificial chess player was demonstrated a little more than 15 years after the advent of the digital computer (Kotok 1962). The social structures of the mid century also venerated professional activities such as medical diagnosis or legal advice giving as feats demonstrating great intelligence, and just as it had seemed that chess should be a grand challenge problem, so it was the case that these activities became a focus of AI research. Again, great initial success was achieved; Mycin (Shortliffe 1974) and Dendral (Buchanan and Lederberg 1971) are examples of so-called 'expert systems' where AI researchers created computer systems with better performance in the real worlds of diagnosis and engineering than any available human expert.

These achievements, the brilliance which characterized the debate on AI, involving some of the greatest thinkers of the late 20th century (for example, Alan Turing, Roger Penrose, Herb Simon), and the fundamental human attraction of a thinking machine

ICT Futures: Delivering Pervasive, Real-time and Secure Services
Edited by Paul Warren, John Davies and David Brown
© 2008 John Wiley & Sons, Ltd

beg a question. If AI is such a successful field of research, why hasn't the vision of a robot in every home come to pass? Where is the world of super intelligent machines that seemed to be offered by the pioneers of AI? More prosaically for companies, like BT, why is it that AI hasn't revolutionized our business?

Some of the answers to these questions lie in a report commissioned by the British Government, the Lighthill Report. In 1973, James Lighthill was asked to examine the prospects and value of research on AI being done in the UK. He delivered an analysis of AI which foreshadowed its subsequent history and took part in a televised debate (Lighthill 1973) where his views were subject to criticism from leaders in the AI community, but emerged relatively unchallenged.

Lighthill describes activity in AI at the time in three categories: the study of industrial automation, the study of human or animal cognition, and the construction of generally intelligent machines. This final category is attacked by Lighthill as useless in the context of his time; he regarded the work that was being conducted in this area as little better than sorcery, and his report led to the withdrawal of virtually all funding for AI research in the UK.

Lighthill's critique is easily summarized. During the early years of AI's development, it had appeared that significant steps had been made toward solving the four key challenges that McCarthy had outlined.

These were

- dealing with the combinatorial explosion of possibilities that searching through spaces like the possible moves that a chess player contemplates opens up;
- representing information internally in the machine in such a way as to be able to manipulate and retrieve it in the same way that humans manipulate and use their experiences;
- inventing ways to instruct or advise the machine so as to bridge the gap between it and the humans that interact with it;
- automatic programming: creating ways for the machine to decide what to do next, rather than following a preset set of instructions.

But Lighthill concluded that it was possible to construct an argument that the achievements of that period were illusory and shallow. Indeed, McCarthy himself acknowledged this in 1974 (McCarthy 1974), and a number of other *mea culpa* style admissions of failure can be found in the contemporary literature (for example, McDermott 1976).

There isn't the space in this essay to examine the failings of AI in all four categories, but taking the first challenge, dealing with combinatorial search as an example, we can use the history of Computer Chess to illustrate the situation that Lighthill discovered.

The reason for 'doing' computer chess was that in the solution to chess and in the journey of discovery to that solution, fundamental mechanisms of general intelligence would be revealed, but this proved not to be the case. The difficulty with building a computer to play chess is that it must either find a calculus that deduces the next move to make from the current position, or it must find the best move to make next by trial and error. There has never been any suggestion of a mechanism that would deduce the next move logically, and so AI researchers focused on the problem of managing the search through the vast number of possible moves that a player can make at any time.

In order to evaluate a move properly, one must not only see what will happen when you make that move (for example, taking the opponent's queen) but what the next move that your opponent would then make would be (for example, putting your king in checkmate). The move that gives you the most instant utility (a queen is the most valuable piece to take) is the worst move to make in some cases (checkmate is the end of the game). Given this, you have to look forward many moves to be able to prove that the move that you are to select is the right one. The number of moves possible grows combinatorially with every move forward that you look, and controlling this combinatorial explosion is the technique that could be used to create AI chess players.

Relatively early on in their work, computer chess researchers were successful in finding general techniques, for example, Alpha-Beta search (see Russell and Norvig 2003, but widely attributed to McCarthy in the 1960s), which dealt with some aspects of the search explosion. However, subsequent innovations were completely domain dependent, such as the use of endgame databases to deal with search problems that were tractable to humans, but were intractable to machines. The history of computer chess is summarized by the incident in 1973 when Donald Michie, Seymour Papert and John McCarthy bet David Levy (an international Chess Master) $3000 that he would be defeated by a computer chess program before the end of 1979; Levy won the bet, and the pioneers of AI were defeated. There was limited incremental progress in the field over the next 15 years as more computing power became available until in the mid 1990s, IBM decided that the time was right for an investment in a high performance computing project using computer chess as a test bed, and the Deep Blue machine was born. This finally beat Gary Kasparov who is widely agreed to be the strongest human chess player. Deep Blue solves the key problem of dealing with search in chess using brute force and not insight. Deep Blue has very little facility for chess, or understanding of it, and it uses very few techniques to control the searches it needs to find wining positions – it's just very, very fast.

Just as computer chess proved to be a brittle model of general intelligence, expert systems proved to be a brittle model of business intelligence. The problem was that while a system with competence in a limited domain could be developed, the limitations of it's competence were hard for its designers to delineate, and it was hard to extend and maintain the system in the face of the dynamism common in the real world. As Lighthill repeatedly said, without theories of intelligence, general technical results would not follow. Application orientated research and research that investigated aspects of cognition in an attempt to formulate theories were respectable activities, but the 'robotic agenda' that many AI departments were pursuing hit a dead end.

12.2 The AI Winter and AI Spring

Together the failure of the AI research program in the middle century either to deliver fundamental science or to provide workable technology resulted in a rapid constriction of funds channeled to AI by both the scientific research bodies that were interested in supporting the philosophical, mathematical and psychological basic research, and the industrial sources interested in the commercial potential of AI technology. Without funding, activity stalled and many people lost interest. The AI community refers to this period – roughly 1973–1982 – as the 'AI Winter', a time in which it was forced to focus on rapid exploitation of known results and during which a framework of theory and practice that would underpin future efforts was quietly constructed.

The Japanese Fifth Generation Computing Project founded in 1982 signaled an unfreezing of progress in AI. At the centers of research which had continued work from 1973 to 1982, progress had been made in developing the tools and models which would enable genuine progress and scientific method. The Fifth Generation project focused on logical inference as the key mechanism for bringing computing interfaces closer to the understanding of normal citizens as opposed to computer scientists, and because of this, it had a strong AI flavor. Indeed, the key implementation language chosen for the Fifth Generation by Japan was the Prolog language developed at the University of Edinburgh. In the UK, the British government instituted the Alvey program in response to a report by John Alvey, a director of technology at BT, and the European Union (EU) started the ESPRIT program. A debate was held at the American Association for Artificial Intelligence (AAAI) conference in 1984 on the topic of an AI winter, but by then, funding was available for AI research from a number of sources worldwide.

The burst of activity prompted by Alvey and the Fifth Generation tapered out in the early 1990s without making the impact that their sponsors had anticipated (Mehta 1990), but the prohibition on AI research had been broken and activity continued, albeit at a lower level than during the Alvey and Fifth generation era.

It was during this time that the key work that has delivered real value from the AI community into society and commerce was done.

12.3 Old Successes and New Failures?

During the 1990s, there were three advances that have allowed the AI community to begin to make a real contribution to our lives.

Firstly, the technology environment that AI operates within has changed out of all recognition. Computers are now amongst the cheapest and commonest scientific instruments, whereas once they were amongst the rarest and most expensive. The arrival of the Internet and the impact that the personal computer revolution and the World Wide Web has had on society have provided AI researchers with completely new test cases and insights into the general problems of intelligence, and indeed applications that matter to people everyday and would be enabled by the development of techniques that could deliver general intelligence. Beyond the change in the environment, the impact on science and technology of information technology has opened the wallets of the research funding bodies to grant applications that promise to replicate this in the future.

The next factor has been the refinement of the working practices of AI which enabled AI researchers to investigate their field more systematically and scientifically. The development of an infrastructure of publications, editorial boards and conferences has led to a wide understanding of the criteria for peer review of AI work. A research culture has developed over the years which enables AI researchers to asses the quality of each others' efforts and to confidently reuse results from project to project. While the criteria and standards of AI research do not match those of the physical or medical sciences (in my opinion at least), they do represent a significant advance on the situation that led to the attacks of Lighthill, McDermott, Minsky and McCarthy in the 1970s and 1980s.

The final factor is the development of several families of algorithms in the AI community which have delivered economically significant results. The techniques of heuristic search, constraint satisfaction, fuzzy logic and data analytics, which have become commercially viable and widely used in industrial applications, were all conceived in the AI boom during the middle century, but bore fruit only after a prolonged period of maturation in the 1980s and 1990s. These three technologies are not the only focus of work in the AI community, but they are the techniques that have proved to be realistic solutions in domains which generate business cases and investment capital.

All of these factors have made AI less prominent in the public mind. Scientific credibility mitigates the hype in the community; the general awareness of and impact of information technology make AI often seem to be no big deal, and the delivery of specific applications focus around the applications themselves rather than the wider topic. But while these successes are not widely promoted as evidence of AI all around us, they are real evidence of its success. As such, it is worth documenting the application of these technologies at BT in slightly more detail as specific examples of how they have impacted a particular company in the recent past. These applications are probably not widely discussed outside BT and its immediate suppliers or peer companies, and so it's a function of this essay to promote them a little more, but there are a number of well-known success stories for AI which have had geopolitical and socially significant impacts, and these are discussed in the subsequent part of the essay to illustrate the wider impact of the field.

12.4 Impact in BT: Heuristic Search and Constraint Satisfaction

As on example of industrial application of AI, BT has developed a range of software applications for workforce management that are underpinned by advanced algorithms which utilize the techniques of heuristic search (Voudouris *et al.* 2001, Voudouris *et al.* 2006) and constraint satisfaction (Lesaint *et al.* 2000). These systems are commercially available as part of the Fieldforce Optimisation Suite (FOS) system.

Constraint satisfaction and constraint programming are techniques that researchers in AI have developed to frame problems such as the eight queens problem. This problem is typical of a class of problems which are formulated by listing a number of things (constraints) which must be true for the problem to be solved. In this case, the problem is to place eight queens on a chess board so that none of them will take another one. The constraints can be written

for each queen Qn[1]

$$!\text{takes}(Qn, Qn + 1) \dots !\text{takes}(Qn, Qz),$$

---

[1] !takes(x,y) denotes that piece x is not able to take piece y.

and

$$!takes(Qn, Qn - 1) \ldots !takes(Qn, Qa)$$

where Qa is the first queen in our set, and Qz is the last queen.

Thus, for each queen, check to see that it does not take another queen.

Heuristic search uses pragmatic rules of thumb (heuristics) to control the combinatorial explosion in search spaces that stymie work on computer chess. Because chess is an artificial game, it turns out that rules of thumb are not very useful in controlling the search space that the computer explores. Sacrifices and the importance of tactical position vs. material gain mean that the search space that the algorithms must explore for a chess game is disordered and discontinuous. In contrast though, it turns out that many natural problems have ordered and continuous search spaces that can be safely navigated by using rules of thumb. For example, one might say that if one job is close to a worker in one direction, and another job is close to a worker in another direction, then jobs that are in-between these two might well be close as well. Given this heuristic, we might conclude that these jobs are worth evaluating before any others.

Heuristic search algorithms are often used to solve constraint satisfaction problems. In the case of BT's field management software, the job allocations required to schedule the work of 30,000 engineers are framed in terms of a constraint based scheduling problem and are then solved partially using a heuristic search algorithm (simulated annealing) (Lesaint *et al.* 2003). The use of the FOS components in BT has saved many tens of millions of pounds through improving workforce productivity and managing the deployment of the company's resources more effectively (Azarmi and Georgeff 2003).

12.5 Impact in BT: Fuzzy Logic and Data Analytics

Fuzzy logic was invented by Lofti Zadeh (Zadeh 1965) to allow logics that moved beyond the bivalence of true and false, but admitted imprecise concepts such as warmth and comparative size. It sometimes does not make sense to build applications that (for example) classify people who are 1.65 m tall as admissible, while people who are 1.64 m tall are inadmissible. Sharp classifications of this type can be used for certain applications, but to define people as tall or short with such a crisp cut-off does not make sense in the wider world. In addition, if the person concerned is classified as 'short', then the information that they are close to the boundary is completely lost to our system. Fuzzy logic introduces a method of describing items in terms of the degree to which they belong to categories. So, we can describe something as 0.8 hot and 0.4 cold, and we can apply functions which evaluate this information appropriately.

In BT, the representations offered by fuzzy logic have been coupled with the algorithms of statistical analysis and machine learning to create decision support systems that enable enhanced customer service management. Concepts such as 'customer ready to buy' can be discovered in datasets gathered by monitoring customer interactions with an organisation (for example, billing enquiries or complaints) and then can be used to modify the actions that are taken to serve them (Nauck *et al.* 2000). The enabling technology here is the use of fuzzy logic as a general representation system and the use of algorithms that can summarize the information in very large datasets into an actionable form very rapidly. The application of the technique of fuzzy data analysis has saved BT over 20 million pounds in the last 3 years.

12.6 The Wider Impact of AI

As on example of industrial uptake, BT has benefited from AI technology in the two specific application areas discussed above. It can be argued that the benefits that BT has received from these two areas (there are a large number of other success stories as well) far outstrip the investment that the company has made in AI research in the past. Of course, BT is benefiting from research that has been funded by Defense Advanced Research Projects Agency (DARPA), the EU and national programs like Alvey. There are, in addition, many other success stories that can be pointed to to indicate why AI research is once again seen as respectable, and we briefly discuss some of these below.

The most prominent success of the AI community is the dynamic analysis and replanning tool (DART) logistics planning tool. This was developed by DARPA for use by the US Army in the 1990s. It was used in operation Desert Shield and Desert Storm, and the savings that it generated repaid the totality of DARPA's investment in AI in the previous 30 years (Hedberg 2002). Subsequently, DARPA has been far more open to investing in AI research, and in 2005, the DARPA challenge project resulted in a number of wholly autonomous vehicles navigating across several hundred miles of desert. In November 2007 a number of autonomous vehicles competed in the Urban version of the Challenge and were able to navigate completely autonomously in a real city environment and complete tasks such as parking and merging with traffic without human intervention (DARPA, 2007).

Fuzzy logic has also had societal level impact. For example, VW has manufactured a gearbox which is fitted into its current model range that applies the same set of techniques to enable better decision making by the car to improve performance and fuel consumption. Automatic focus and image shake management features on digital cameras also use fuzzy logic. While camera manufacturers are extremely wary of labeling their products as fuzzy, Canon and Sony use fuzzy logic for autofocus, while Nikon uses it for light metering. AEG, Zanussi and Maytag (amongst others) use fuzzy logic for controlling the behavior of washing machines and other appliances.

12.7 The Future

New and exciting topics for AI research are emerging as part of the ongoing interaction between the new pervasive computing environment, successful application and the scientific methods of AI research that have become dominant in the literature and community in recent years. Important new technology is emerging from the Semantic Web that enables the interface between humans and computers to be easier, and the intelligent agent community has developed technology that promotes the ability of computers to undertake tasks in our daily lives without requiring constant attention or constant correction.

The development of machine minds is being driven in two directions: work that allows machines to model and understand humans, and work that allows them to model and understand the world that they exist within. It is becoming increasingly apparent that the world of machines is the world of the Internet, and that computers will exist within a universe that is fundamentally different in a number of ways from the universe that humans inhabit. The mechanism of thermal decay and change that underpins human notions of time, and the mechanisms of inertia that underpin our notions of space are not pertinent on the Internet, and the conceptual machinery required to understand them is therefore irrelevant there. However, introducing these concepts to the man–machine interface has proved (via desktop metaphor GUI's for example) extremely significant.

I believe that it is in this gap, the chasm of difference between the virtual world and real world, that the real need for AI research in the next decade lies. Internet search engines provide almost everyone with access to vast libraries of data and information, but lack the knowledge and insight of a human teacher. In parallel to the rise of social networking sites like Facebook and MySpace, there has been large growth in the use of video games. Millions of people interact with artificial characters in video games everyday, but these interactions are means to an end; on the other hand, the millions of interactions between subscribers on social networking sites are ends in themselves. Essentially, there is no dialog between humans and machines because while both sides can speak, neither has anything to really communicate to the other.

12.8 The Next Steps in AI

At the beginning of this essay chapter, I asked why it was that AI was not a part of everyday life and business. I hope that this question has been answered by the history of the subject and that the real impact of AI research today has been demonstrated through the prosaic and mundane reality of its use in washing machines, cars and cameras. The benefits of continued advances in both the understanding of general intelligence and the application of this understanding to new applications seem just as likely to continue to flow as the fruits of any other scientific pursuit. There is something that it's important for those interested in AI to understand. Instead of asking the question of what does the future hold for us, I think this is a good time to ask what kind of a future do we want?

Intelligence, consciousness, the mind, call it whatever you want, is by increments being understood. I feel that it's unlikely that its mystery will be revealed in the immediate future, but I think it's clear that it is an artifact of the physical universe that can be understood in the same way that the workings of a star and the motions of the heavens can be understood. What is more, it seems likely that just as the mechanisms that power a star can be recreated in a torus of super heated gas, the workings of the mind can – and probably will be recreated in some machine or other made by men. It may be a digital computer, although there are some arguments that such an infrastructure is inadequate for an intelligence (Penrose 1994), but there is no indication that the creation of some sort of suitable device is physically impossible, although it is fair to admit that it may be quite impractical – we simply do not know.

To date, we have been able to create narrow intelligences with abilities at the level of lower animals such as insects. Our robots can find their way across deserts (if we ask them to), and they can beat us at chess, although while it is clear that the robots understand a lot about desert roads, it is clear that they understand very little about chess. More significantly at the moment, our robots currently help us to schedule our days, change gear, take photos and figure out if the washing is quite done in our hard water with our poor quality soap. We have a framework of practice in which to build theory and record observations and engineering advances for others to build on. While AI might not be quite the science that it aspires to, the significance of that aspiration and the genuine attempt of the community to meet it gives us hope that the long, steep and rocky road that we must stumble up to understand the mind will gradually be conquered. As our competence grows, so will the scope of application that can be tackled. It may be that there is never a moment where a scientist needs to press a button to raise the computer to life, but at some point, we will be able to identify that the threshold of personal, self-reflective and feeling intelligence will be crossed.

Do we want to do this? In popular science and in science fiction, a human level AI is often characterized as potentially malign, but the reality is that any intelligence that emerges or is created will be the child of our mind, and therefore surely our anxieties are simply the fears that we all hold about the potential for wickedness in any child. The real concern though should not be for humans, because while we are wonderful creatures, surely we have no greater significance than other higher mammals. Instead, as always, we should be concerned for the other; we should care about the welfare of anything that we bring into this world.

What kind of life would it be for a machine brought to consciousness as part of a high-tech freak show? Do we have the mercy in us to refrain from inflicting such a trial on an equal? Will we know what to say when it wants to know what we created it for? Will we be able to face our creator having done such a thing?

References

Azarmi, N. and Georgeff, M. (2003) 'What Has AI Done for Us.' BTTJ, 21(4), 15–22.

Buchanan, B. and Lederberg, J. (1971) 'The Heuristic DENDRAL Program for Explaining Empirical Data.' Stanford University, Stanford, CA.

DARPA (2007) Defense Advanced Research Projects Agency, Grand Challenge 2007. Available from: http://www.darpa.mil/grandchallenge/docs/Urban_Challenge_Participants_Conference_FINAL.pdf, downloaded 1 October 2007.

Hedberg, S. (2002) 'DART: Revolutionizing Logistics Planning.' IEEE Intelligent Systems, 17(3): 81–83.

Kotok, A. (1962). A chess playing program for the IBM 7090. Massachusetts Institute of Technology. Department of Electrical Engineering.

Lesaint, D., Voudouris, C. and Azarmi, N. (2000). 'Dynamic Workforce Scheduling for British Telecom plc.' INTERFACES, 30(1): 45–46.

Lighthill, J. (1973) BBC TV – June 1973 'Debate at the Royal Institution.' Available from: http://www.aiai. ed.ac.uk/~dm/, downloaded 1 October 2007.

McCarthy, J. (1974) 'Professor Sir James Lighthill, FRS. Artificial Intelligence: A General Survey.' Review in: Artificial Intelligence, 5(3): 317–322.

McDermott, D. (1976) Artificial Intelligence meets Natural Stupidity. ACM SIGART Bulletin, 57: 4–9.

Mehta, A. (1990) 'Ailing after Alvey', New Scientist, 7 July 1990.

Nauck, D., Spott, M. and Azvine, B. (2003) 'SPIDA – A novel Data Analysis Tool.' BT Technology Journal, 21(4).

Penrose, R. (1994) 'Shadows of the Mind: a Search for the Missing Science of Consciousness.' Oxford University Press.

Russell, S. and Norvig, P. (2003) 'Artificial Intelligence: A Modern Approach, Second Edition.' Prentice Hall.

Shortliffe, E. (1974) MYCIN: A rule-based computer program for advising physicians regarding antimicrobial therapy selection. Ph.D. dissertation, Stanford University.

Voudouris, C., Dorne, R., Lesaint, D. and Liret, A. (2001) 'iOpt: A Software Toolkit for Heuristic Search Methods.' In T. Walsh (ed.) *Principles and Practice of Constaint Programming – CP 2001*. Lecture Notes in Computer Science vol. 2239, 716–729. Springer.

Vodouris, C., Owusu, G., Dorne, R., Ladde, C. and Virginas, B. (2006) 'ARMS: An Automated Resource Management System for British Telecommunications plc.' *European Journal of Operational Research*, 171(3): 951–961.

Zadeh, L. (1965) 'Fuzzy Sets.' *Information and Control*, 8.

Part Three

Applying Technology

13

Healthcare

Chris Wroe, Paul Garner and Janette Bennett

13.1 Introduction to the Healthcare IT Setting

Every organisation, industry and service experiences internal and external pressures that result in change – from small evolutionary steps to revolutionary. Internationally, healthcare has arrived at a crossroads with all routes pointing to the need to utilise technology in all aspects of delivery to meet demands. Healthcare costs are escalating exponentially due to, amongst others, expensive drugs, designer therapies, consumer demand and chronic lifestyle diseases (World Economic Forum 2007). Technology is therefore needed to help patients to help themselves. In another direction, there is a known healthcare divide (Department of Health 2007), with high standards of care and outcomes for the privileged and an increasingly lower standard of care for the rest. Technology is needed to enable those who are hard to reach to gain the same level of access. Awareness of the degree of iatrogenic[1] and organisationally induced injury (National Patient Safety Agency 2007) and associated medico legal costs have created a consumer, clinician and, political demand for improved safety, so decision support tools are in demand not just to support the clinician, but to track variation in their compliance. All at a time when staffing resources are reaching a crisis (Perlino 2006), the number of informal carers is rising while the number of tax payers decrease faster than the number of retired persons (Costello 2002, George 2001). Healthcare educationalists and policy makers are left wondering what the learning and provision needs of the next generation of care providers are, and even who or what those providers may be. Healthcare professionals providing care are also at a loss as to how best to deliver evidence based medicine – each year, Medline indexes 560,000 new articles, and Cochrane Central adds 20,000 new trials per year (this is about 55 per day), an impossible volume for any clinican to keep up to date with.

---

[1] i.e. resulting from medical treatment.

ICT Futures: Delivering Pervasive, Real-time and Secure Services
Edited by Paul Warren, John Davies and David Brown
© 2008 John Wiley & Sons, Ltd

In short, new ways of delivering healthcare is essential for economic stability and for maximising clinical capability (Gingrich *et al.* 2006). As Wanless (2002) suggests, citizens must have, and take, more control over their healthcare, and more investment in technology is needed to support clinicians in coping with the reduced availability of resources, and to make education more responsive to an evolving and unknown demand. But Coiera (2004) points out there are four rules to be taken into account for the reinvention of healthcare:

1. Technical systems have social consequences.
2. Social systems have technical consequences.
3. We do not design technology; we design sociotechnical systems.
4. To design sociotechnical systems, we must understand how people and technologies interact.

13.2 Telemedicine, Telehealth, and Telecare

Just as the terms clinical and medical are often used interchangeably, so are tele and ehealthcare, and telematics. For the purposes of this chapter, we will use the following, taken from Wikipedia (2007), itself a new way of developing and using educational resources.

The term *Telemedicine* is the delivery of medicine at a distance. The term is composed of the Greek word τελε (*tele*) meaning 'far', and *medicine*. Telemedicine may be as simple as two health professionals discussing a case over the telephone, or as complex as using satellite technology and videoconferencing equipment to conduct a real-time consultation between medical specialists in two different countries. It can also involve the use of an unmanned robot.

Telehealth is the delivery of health related services and information via telecommunications technologies. It, too, may be as simple as two health professionals discussing a case over the telephone, or as sophisticated as using satellite technology to broadcast a consultation.

Telecare is the term given to offering remote care of elderly and vulnerable people, providing care and reassurance needed to allow them to remain living in their own homes. Use of sensors allows the management of risk and as part of a package which can support people with dementia, people at risk of falling or at risk of violence, thereby preventing hospital admission.

eHealth (also written *e-health*) is a relatively recent term for healthcare practice which is supported by electronic processes and communication. The term is inconsistently used: some would argue it is interchangeable with healthcare informatics, while others use it in the narrower sense of healthcare practice using the Internet.

In all cases, services themselves are either synchronous or asynchronous:

Synchronous involves real-time systems to monitor specifics such as vital signs using sensors or videophones that communicate with a remote call centre or base. Responses to signals received may be an automated alarm or telephone call to a carer or an alert to escalation or emergency services.

Asynchronous is often used for self-care, using personal digital assistants (PDAs), personal computers (PCs), and mobile phones which may be either wireless or hard wired. Examples are the recording of personal test results – blood sugar or blood pressure. These can communicate with remote call centres or can be home based, and both provide and receive data for clinician review or feedback to the end user.

There are several types of connections used with real-time exchanges. Standard analog telephone lines can support videophones and include a camera, display screen, and telephone. Because these are available in most Western homes, they are the easiest to set up. It is also possible to use a television (Camarinha-Matos and Afsarmanesh 2004), but although resolution is better than videophones, not all televisions have the necessary connectors.

Virtual reality is one of the newest tools available. The computer technology, consisting of both hardware and software, allows three-dimensional (3-D) virtual environments. The current popular, technical, and scientific interest in virtual environments is inspired, in large part, by the advent and availability of increasingly powerful and affordable visually oriented, interactive, graphical display systems and techniques.

13.2.1 Telemedicine in Use

In the USA, the Veterans Administration actively uses telemedicine for people with disabilities, providing physical examinations, monitoring, and consultation for veterans with spinal cord injuries. A few school districts in Oklahoma and Hawaii offer school-based rehabilitation therapy using therapy assistants who are directed by a remote therapist. The National Rehabilitation Hospital in Washington DC and Sister Kenny Rehabilitation Institute in Minneapolis also provided assessment and evaluations to patients living in Guam and American Samoa. Cases included post-stroke, post-polio, autism, and wheel-chair fitting.

In 2001, O. Bracy, a neuropsychologist, introduced the first web based, rich Internet application, for cognitive rehabilitation therapy. This system provides, directly to the patient, the therapy prescription set up and controlled by the member clinician. All applications and response data are transported via the Internet in real time. Patients can log in to do their therapy from anywhere they have access to an Internet computer. In 2006, this system formed the basis of a new system designed as a cognitive skill enhancement program for school children with individual children or whole classrooms participating over the Internet. More recently in the UK, the Department of Health mandated that this type of facility must be available to patients with depression.

Applications have also been developed to assess and/or treat acquired adult speech and language disorders, stuttering, voice disorders, speech disorders in children, and swallowing dysfunction. Recent applications have involved the use of sophisticated Internet-based videoconferencing systems with dedicated software which enable the assessment of language disorders (Georgeadis *et al.* 2004) and the assessment and treatment of motor speech following brain impairment and Parkinson's disease disorders (Hill *et al.* 2006). Collectively, these studies have revealed positive treatment outcomes, while assessment and diagnoses have been found to be comparable to face-to-face evaluations. A sophisticated telerehabilitation application for the assessment of swallowing was developed by Perlman and Witthawaskul (2002) who described the use of real-time videofluoroscopic examination via the Internet. This system enabled the capture and display of images in real-time with only a 3–5 second delay.

In many countries, patients who would normally be cared for in acute care facilities with problems such as heart failure are now being managed at home (Johnson 2006). Patient physiological monitoring data are received by a centralised data centre providing technical support and data backup, with acute care clinicians reviewing the data collected.

13.2.2 Telehealth in Use

In 2005, at least one third of the European adult population, 130 million European Union (EU) citizens, browsed the web in search of information on health. To help European citizens answer their health questions, the Commission launched the 'Health-EU Portal', a gateway to simple and sound information on 47 topics that range from babies' health to bio-terrorism, and from infectious diseases to health insurance (Health-EU Portal). Web surfers will have access to over 40,000 links to trustworthy sources. The translation of the portal into all 20 official EU languages means that up to 1.5 billion people worldwide can use it.

One of the main goals of the portal is to help people take responsibility for and improve their own health. It provides information on a wide range of health concerns. Forty-seven topics are divided into six thematic areas:

- 'My Health' – e.g. women's health, people with disabilities, babies and children (nutritional advice, toy safety tips, etc.);
- 'My Lifestyle' – e.g. nutrition, drugs, tobacco, sports and leisure, travel advice (pan-EU emergency call number, what to do if you get sick in another member state, etc.);
- 'My Environment' – e.g. at home, road safety, consumer rights;
- 'Health Problems' – e.g. cancer, mental health, cardiovascular diseases (nutritional and lifestyle advice, etc.);
- 'Care for Me' – e.g. long-term care, insurance, mobility, medicines;
- 'Health in the EU' – e.g. research, indicators, statistics.

In the UK, NHS Direct provides a 24 hour nurse-led telephone advice service run by the NHS which not only provides information on the diagnosis and treatment of common conditions but also runs http://www.nhsdirect.nhs.uk/, a website providing health information.

To remain independent, and use less of scarce clinical resources, citizens who become patients need not just education and information but also:

- instructions, often embedded in care plans;
- assistance in managing compliance and concordance with drug regimes (those with long-term conditions and increasing age can have over 10 drugs to manage daily);
- the ability to monitor their own health status;
- feedback on how they are managing;
- ready access to a care team.

This last area is in the domain of the patient's home, outside of hospital care, and overlaps significantly with social care.

13.2.3 Telecare in Use

Telecare is the use of information and communication technology (ICT) to provide support to enable vulnerable people to live independently at home. This encompasses a whole spectrum of technologies, which can be divided into three generations:

1. 1st generation telecare systems are typically panic alarm systems which are activated by the individual in an emergency. Usually, these are a pendant or wristwatch worn by the client, or a pull cord situated somewhere in the home.
2. 2nd generation telecare systems use sensors in the home to monitor clients and/or their surroundings, raising an alert if an unusual or dangerous situation is detected. They differ from 1st generation systems in that the client does not have to actively trigger the alert. These can include movement and fall detectors, temperature sensors, and flood detectors.
3. 3rd generation telecare systems enable longitudinal monitoring of a person's well-being, detecting deteriorations and improvements. It is proposed that these systems could be used in care assessment processes or to measure the effectiveness of a care package for an individual.

Alarms are raised in response to situations such as unusual lack of activity which is a departure from the client's normal behaviour pattern; abnormal usage of the fridge, cooker, bath, toilet; property not being secure at night; hazard warnings such as low room temperature. West Lothian's 'Opening Doors for Older People' trial is recognised as one of the most advanced social healthcare projects in Europe (Gillies 2001). However, Liverpool has also implemented a telecare system implemented by BT which does not require the client to wear or carry any devices, nor does the client have to interact with a complex user interface. The intelligent analysis within the system removes the requirement for clients to initiate a call for help themselves and instead varies the automated alert time thresholds, by time of day and room, slowly over weeks, to optimal values based on the personal behaviour of the individual.

The system automatically determines time thresholds for 'lack of activity' and 'lack of room change' for an individual client and exhibits the following key features:

- The algorithms are adaptive and automatically calculate new alert thresholds for the individual based on a rolling window of their activity levels over the past 3 months (decayed by time).
- The algorithms can provide separate thresholds for differing room types within the home, living room, hall, bathroom, bedroom, and kitchen, so accommodating the differing activity levels in each.
- The thresholds can be further subdivided into 4 hour time periods, allowing the differing behaviours of the individual throughout the day to be accommodated.
- The algorithms have a unique feature that allows care provider policy to be embodied within the automatic determination of thresholds.

The efficacy of these systems is becoming more evident as projects go beyond pilot investigation (Hensel *et al.* 2006). Smart homes are not exclusive to the UK. Ubiquitous computing and home based technologies are becoming a reality in many countries.

13.2.4 Pervasive Telecare

Telecare systems for monitoring people at home have so far focused on ensuring safety and security of individuals living alone in the community, so-called reactive or r-mode solutions. The r-mode solutions are based on well-established technologies with proven reliability and effectiveness. New technologies, or perhaps more importantly, the convergence of technologies, may enable a new suite of services for individuals which go far beyond r-mode solutions. Pervasive computing is one such technology trend, which, when combined with intelligent networked services, could have a significant impact on health and social care services, enabling smart personalised solutions tailored to the individual and focused on quality of life and well-being. Such solutions offer the promise of optimised care plans, integrated quality management of service provision, and a movement to preventative or p-mode services. An early prototype of pervasive computing in the care sector was developed by BT, Liverpool City Council, and the Universities of Dundee, Liverpool, Loughborough, and Bristol collaborating under UK Department of Trade and Industry sponsorship. The collaboration culminated in field trials with two Social Services clients whose homes were equipped with up to 40 sensors in order to track changes in their well-being. The sensors detected a wide range of functions including movement, furniture occupation, door movement, general use of tables and appliances, use of water, electrical power, entertainment and communications devices. Three aspects of client activity were monitored: long-term trends, significant patterns, and associations amongst patterns. All the substantial or gradual changes in daily activities were reported to the carer using a PDA, with the detailed analysis conveyed in a way that allows the non-IT expert to easily understand the results, available through a PC interface. Subtle changes in human activity levels, related to well-being, were measurable on a continuous objective basis. The sensors used were all unobtrusive, battery powered, and did not include video, audio, or tag based devices to protect the privacy and dignity of the client.

In attempting to track changes in health and well-being before crises occur, the richest possible set of data relating to the individual client is required. Individuals with some chronic diseases can now be supported through management of data from devices which measure a specific physiology, in order to provide personalised health information and targeted clinical interventions if required. Combining activity monitoring with on-body physiology monitoring will further expand the range of needs that can be met by telecare systems. Combined systems will allow physiological data to be put into its activity based context and thereby enhance its value.

Pervasive computing enabled solutions are likely to provide greater insight into the effectiveness of prescribed drugs, medical treatment, and social care interventions, offering the promise of optimised care services, tailored to the individual needs of each user, based on continuous objective evidence. Integration of secure and trustworthy pervasive home monitoring with converged voice and data services across multiple health and social care suppliers underpins our vision of a responsive, flexible, affordable service able to provide personalised preventative care to those who need it.

13.3 Communication Technology and Advancing Practice

In fulfilling their duty of care to their patient, clinicians must coordinate and refer to other healthcare professionals, and develop the knowledge of other clinicians. All of these activities are being facilitated by communications technology.

The use of teleconferencing for multidisciplinary team meetings (MDTs) is increasingly common across the UK, especially in the areas of cancer. The meetings are used to ensure seamless delivery of care throughout treatment across the boundaries of primary, secondary, and tertiary care. 'Virtual' MDTs also use video conferencing to share knowledge and expertise that may not be available locally.

Some organisations are increasing providing clinical educational activities for all health professionals who live and work in remote locations. Scotland is one such area providing education services to four regions: Grampian, Highland, Orkney, and Shetland. The project supports videoconference sites which interconnect via a Tandberg 'bridge' which enables up to 16 sites to link together for a single interactive teaching session or three smaller simultaneous training sessions (Swan and Godden 2005). The bridge also enables participants to join the training from a mobile or standard phone and videoconference units. The equipment ensures high quality transmission of various educational materials, for example, PowerPoint presentations, videoclips, X-ray images, etc., to and from remote locations. Use of this technology is not limited to the Western world. In 2006, World Health Organization launched its knowledge gateway initiative (IBP initiative 2007), linking family nurse practitioners in the field globally to share information and best practice.

13.3.1 International Impact

The shortage of healthcare workers in the Western world led to overseas recruitment until the negative impact that this had on the less fortunate was recognised. Recruiting from Africa left services unmanned with little hope of reestablishing them (Duff 2006). Now the use of technology, and improved standards in interoperability, and the ability to maintain semantic meaning, through the use of Snomed CT, has meant that the shortage can provide other countries with an income opportunity as well as retaining its own skilled staff. For some years now, radiology reports have been provided to the USA and Denmark using radiographers in India with Australia too looking to India to provide much needed clinical resources for its own citizens but without the need for relocation (Ghanta 2005).

13.3.2 Career Advances

Decision making by carers requires access to knowledge support. Increasingly, there is a demand for health service 'brokers' or clinical shepherds who help consumers navigate healthcare systems and knowledge resources when making their own choices on care. Here, communication technology accessing knowledge support systems are essential.

13.3.3 Self-care Advances

Our children already use Internet services such as MySpace for peer support on latest issues of interest to them. In September 2006, MySpace had 106 million accounts with a further 230 thousand registrations per day.

Both young people and adults also display a strong propensity for social networking online, e.g. Friends Reunited, and even for the formation of meaningful relationships with machines, as with Tamagotchi toys.

Building on this, there is a new generation of teleconferencing capability and the potential for the creation of 'Second Life' (http://www.secondlife.com/) type virtual worlds in which medical/health related exchanges could occur with peers, fellow sufferers, and clinicians. Your support networks (including clinicians) do not have to be physically present (not even to touch you!). Research into the use of avatars to allow individual health modelling is already starting.

Ground-breaking research in telehaptics (the sense of touch) (Seelig *et al.* 2004) and virtual reality may broaden the scope of telemedicine, in the future. As already discussed, it is possible to tele-mentor so that an experienced surgeon or specialist can use a two-way video conferencing link to guide a lay person or doctor at a remote location through an operation or procedure. Tele-robotics uses virtual-reality technology and robotics to allow a surgeon to remotely operate on a patient from hundreds or thousands of miles away (NASA 2005). In both cases, a highly reliable and high-performance network is crucial for conveying the life-saving information and dependably operating sensitive robotic surgical equipment (Waltner 2004).

13.4 Information System in Healthcare: Advancing Care Delivery

'The future has already happened, it's just not very well distributed.' (William Gibson)

There is much that can be done today just by distributing knowledge and practice using existing ICT. However, the world of the care professional is changing.

13.4.1 Patient Record Systems

Internationally, the role of the doctor within the domains of primary and secondary care is largely the same. Their information needs could be assumed to be stable and served by stable and separate systems. Indeed, currently, much practice is supported by paper based systems.

However, other professional groups have different and evolving roles and responsibilities according to the educational processes and health organisation responsibilities of each nation. For example, the UK and USA have nurse and therapy consultants, whereas many other nations do not.

Also, within the UK and the USA, even primary care doctors, also known as general practitioners (GPs), are adopting new roles. In the UK, GPs with special interests are providing services normally associated with secondary care doctors, with the service overseen by a lead consultant.

This is enabled both by ICT use in education and by the ability to share patient's electronic care records and physiological monitoring data using newly developed IT systems. Completely new roles are also developing, made possible by new technology.

In the UK, emergency care practitioners will provide a new service to that of paramedics in that they will not only respond to emergency call outs, but will be able to book the patient a visit with their GP or a hospital appointment in addition to treating them in their own home. The primary intention is to proactively monitor those patients who are known to have acute exacerbations of a chronic disease and thus to avoid hospital admission by being able to deliver care in the patient's own home, supported by access to electronic care records. This will only be possible through the ability to share records between organisations, using access controls. This is all enabled though wireless working as part of the English National Health Service's National Programme for IT (NPFIT). Even complex data can be sent to an expert source, evaluated, and diagnosis received without moving the patient unnecessarily, or moving to a suboptimal health organisation for the patient's condition

13.4.2 Geographic Information Systems

The use of geographic information systems can also support the practitioner through:

- automatic: positioning technology allowing real-time data capture of location based events direct from the field, e.g., home address;
- security: ability to track practitioners in high-risk environments;
- visit support: address finder, route planner.

Organisations, professions, and roles as we know them will change with the emphasis on enhancing and realising the knowledge and diagnostic skill of the professional, rather than the concentration on task. As a result, the near future is one in which clinicians do not carry out any activities that patients can safely do for themselves, or with the support of nonhumans – robots or their own personal, and human nonprofessional carers.

13.4.3 Clinical Decision Support Systems (CDSS)

To provide immediate care, many CDSS and tools are already in use supporting activities that are:

- administrative: supporting clinical coding and documentation, authorization of procedures, and referrals;
- managing clinical complexity and details: keeping patients on research and chemotherapy protocols; tracking orders, referral follow-up, and preventive care;

- cost control: monitoring medication orders, avoiding duplicate or unnecessary tests;
- clinical diagnosis and treatment plan processes, and promoting use of best practices, condition-specific guidelines, complex algorithms for dose calculation, or help to determine probable diagnosis and population-based management.

Many tools cut across all of these categories, such as picture archiving and communication systems [National PACS Team (NPFIT) and National Radiology Service Improvement Team 2005]. Expert systems are now in routine use in acute care settings, clinical laboratories, educational institutions, and incorporated into electronic medical record systems – such as the GP patient electronic record solutions used in the UK. Others are largely stand alone advisory solutions such as ISABEL (Ramnarayan *et al.* 2004). Some CDSS systems have the capacity to learn, leading to the discovery of new phenomena and the creation of medical knowledge. These machine learning systems can be used to develop the knowledge bases used by expert systems, assist in the design of new drugs, advance research in the development of pathophysiological models from experimental data (Coiera 2003). Benefits of CDSS include improved patient safety, improved quality of care, and improved efficiency in healthcare delivery.

13.5 Standards

Throughout this chapter, we describe how healthcare is changing, using technology to cut across existing organisational and geographic boundaries. Existing health information systems are rooted within organisations or even more commonly departments within organisations. In order to keep pace with changing healthcare, information must flow to follow the patient (or further), overcoming technical and organisational challenges. Such technical challenges include data in different proprietary formats, different structures, and are expressed using different vocabularies. Organisational challenges, while not dealt with further here, are equally important such as respecting privacy while understanding when it is appropriate to share information.

Common health IT standards are seen as a foundation for overcoming technical challenges to integration. Two examples will be described here: *SNOMED-CT*, a standardised vocabulary of clinical terms, and Health Level 7 (*HL7*), a set of specifications for standardised messages between systems. Other standards also exist such as *DICOM*, a set of specification for the exchange of medical image and associated data.

SNOMED-CT is the Systemized Nomenclature of Medicine – Clinical Terms. It was formed by the merger of a US medical terminology *SNOMED* with a UK medical terminology *Clinical Terms*. It aims to support the recording of clinical information using a controlled vocabulary that then enables machine interpretation whether simply for information exchange, or for decision support, aggregation, and analysis. Its ongoing development is overseen by the International Health Terminology Standards Development Organisation (IHTSDO).

SNOMED-CT is concept based. A health concept can be represented by more than one term either in one language or in different languages. For example, 'pneumonia' is the English term for a health concept also referred to with the Spanish term 'Inflamación Pulmonar' allowing flow of coded information across language boundaries.

Machine interpretable definitions for concepts within SNOMED-CT, such as 'pneumonia' is a 'lung disease', provide a common terminology resource with which to build upon in healthcare software applications such as clinical decision support. This facilitates the preservation of data meaning from one IT system to another, and holds out the promise of reducing the current effort of developing a bespoke vocabulary within each application.

Standardised vocabulary is not enough but must sit in standardised information models or data structures. HL7 is a standards organisation which develops message specifications to enable consistent exchange of information between healthcare applications (Jones and Mead 2005). The specifications define overall information models that the data transmitted must conform to, interaction patterns defining when messages are sent within a healthcare process, the structure of the messages themselves and the standard vocabularies to be used for each item. HL7 version 3 is the latest to be released and includes support for SNOMED-CT described above.

The existence of standards does not itself guarantee interoperability. They must be successfully implemented. Indeed, these standards are being applied in a variety of healthcare IT settings to support greater interoperability. Within the UK, the NPFIT within the National Health Service is developing national and local care record systems that bring together clinical information about a patient from disparate sources (NHS Connecting for Health)[2].The specifications for how this information is transferred depend heavily on HL7 and SNOMED-CT. On an international scale, a consortium called 'Integrating the Healthcare Enterprise' (IHE)[3] is creating a process that brings together standards such as those described above to address specific requirements, then develops technical guidelines that manufacturers can implement. IHE then stages 'connectathons' which vendors attend to test and demonstrate the interoperability of their products. These integration initiatives are not confined to vendors of proprietary software. The Open Healthcare Framework[4] is a project within the Eclipse open source community formed for the purpose of promoting the development of health informatics technology. The project aims to develop a set of extensible frameworks which emphasize the use of standards to enable interoperable open source infrastructure, and so reduce technical barriers from the bottom up. Not surprisingly, there is significant cross fertilisation between this and organisations such as HL7 and IHE.

The complexity of the health information that needs to be exchanged means that significant challenges still remain to realise the goal of interoperability where a person's health information straightforwardly follows their journey around and between care organisations. Existing healthcare applications implement proprietary data models that predate models developed as part of the standardisation process. Also, in many cases, the line is blurred between where one standard begins and another ends. For example, the information models underpinning SNOMED-CT and HL7 version 3 overlap in a significant number of areas. To address these challenges, researchers are beginning to look at the suite of technologies associated with the Semantic Web.

---

[2] http://www.connectingforhealth.nhs.uk.
[3] http://ihe.net.
[4] http://www.eclipse.org/oht/.

In the words of the WorldWideWeb Consortium,

'The **Semantic Web** provides a common framework that allows **data** to be shared and reused across application, enterprise, and community boundaries. It is a collaborative effort led by W3C with participation from a large number of researchers and industrial partners.' (W3C 2001)

One of the cornerstones of the Semantic Web is the Ontology Web Language (OWL)[5] together with description logic reasoners that can check the consistency of concepts defined in OWL and can also infer new relationships between them. An ontology provides a shared understanding of a domain, in the same way that the models underpinning HL7 and SNOMED-CT seek to do in the healthcare arena. It has been demonstrated that the expressivity of OWL enables HL7 and vocabulary models such as SNOMED-CT to be expressed in the same environment and then checked for consistency using reasoner software (Marley and Rector 2006). Even when confined purely to the representation of vocabularies such as SNOMED-CT, it is becoming apparent that OWL and associated software has a valid place in its development and deployment.

The Semantic Web also includes technology to represent data, the Resource Description Framework (RDF).[6] This provides a flexible graph based model to represent structured data with several advantages over alternative approaches including a standard mechanism for the identification of resources (the Universal Resource Identifier), a mechanism for the aggregation of data from distributed sources, and a link to the well-defined semantics of OWL. It is not clear how or when RDF will find a place in the implementation of health applications (Wroe 2006). However, the Health Care and Life Sciences Interest Group run within the Semantic Web activity at W3C (http://www.w3.org/2001/sw/hcls/) is a focus for active developments and continues to be the resource to track progress in this emerging area.

13.6 Healthcare Devices

The use of remote sensing devices has already been described to support the use of telecare. Devices can also be used to restore function. Implantable devices such as cochlear implants restore a degree of hearing, or prosthetic devices such as artificial limbs allow athletic function to compete with the able-bodied. As this chapter concentrates on supporting the healthcare enterprise rather than advances in medical technology itself, we will focus on the use of robotics in care.

13.6.1 Robotics in Care

A Japanese domestic robot, made by Mitsubishi Heavy Industries, is primarily intended to provide companionship to elderly and disabled people. The robot has two arms, and its flat, circular base has a diameter of 45 cm. The first hundred went on sale in September, 2005, for USD $14,000. Wakamaru runs a Linux operating system on multiple microprocessors. It can connect to the Internet, and has limited speech (in both male

---

[5] http://www.w3.org/2004/OWL/.
[6] http://www.w3.org/RDF/.

and female voices) and speech recognition abilities. Functions include reminding the user to take medicine on time, and calling for help if it suspects something is wrong.

13.6.2 Robotics in Surgery

Few surgical robotic systems have Food and Drug Administration approval. However, the Da Vinci has approval for many surgical procedures including general laparoscopic surgery and laparoscopic radical prostatectomy, thoracoscopic, and thoracoscopically assisted cardiotomy procedures (U.S. Food and Drug Association 2005). It is also used for endoscopic coronary artery bypass graft surgery (Mohr *et al.* 2001). The da Vinci Surgical System is a computer-enhanced minimally invasive surgical system consisting of three components: the InSite® Vision System, Surgical Cart, and Surgeon Console. The InSite Vision System provides the surgeon with a 3-D view of the surgical field including depth of field, magnification, and high resolution. The Surgical Cart includes the EndoWrist® instruments. It has up to four robotic arms whereby the laparoscope and up to three EndoWrist instruments are inserted. The range of motion allows precision that is not available in standard minimally invasive procedures. The Surgeon Console contains the master controls that the surgeon uses to manipulate the EndoWrist instruments.

Robotic intervention is also already possible with New York based surgeons removing a gall bladder in France in 2001 using videoconferencing, telecoms, and verbal commands to instruct (Express Healthcare Management 2001).

The American military since the early 1990s been looking at how to use robotics to avoid sending medical staff into war and disaster zones and yet support those very areas more effectively. While it is currently perceived that robots are still too heavy and expensive to deploy, work continues; the most recent trials of robotic surgery were held undersea as part of NASA's Extreme Environment Mission Operations (Rosen and Hannaford 2006). The wireless technology needed to support this type of activity has so far used geosynchronous satellites, but there is investigation of the use of land launched drone aircraft to support wireless working in the field.

13.7 Future Health IT – Pharmacogenomics and Personalised Medicine

Medicines act differently on individual patients, and under different conditions often interacting with other drugs, diet, and environmental factors. However, an increasing body of evidence suggests that individuals' genetic make-up, their genotype, can play a significant role. Israel *et al.* (2004) found that about one sixth of the American population with a particular genotype had an adverse reaction to a drug commonly used to treat asthma. The ability to undertake large-scale genotyping already exists (Mitchel 2006); each child born in the UK, Australia, and many other countries has a heel prick test (Guthrie test) which detects a number of metabolic genetic diseases by screening a blood sample taken when the baby is under 10 days old. In the future, the genotyping data collected could be useful for a drug that will not be invented for 50 years. When a doctor wants to prescribe a new drug, the information could be checked using decision support systems against the individual genetic make-up to see if it had been shown to be ineffective for the patients' particular genotype.

13.8 Conclusion

The chapter has shown how fast, reliable communications technology that can support novel collaborative services is increasingly vital to delivering efficient and effective healthcare across a distributed community.

These services go far beyond supporting personal communication. Instead, they are becoming proactive. They organise and share care records, provide guidance tailored to individual circumstances, and incorporate devices that can both monitor and manipulate. Development and uptake of such services will continue to gather pace as long as the increasing demands on the healthcare enterprise remain.

References

Camarinha-Matos, L. and Afsarmanesh, H., 2004. *TeleCARE: Collaborative virtual elderly care support communities.* The Journal on Information Technology in Healthcare 2(2): 73–86.

Coiera, E., 2003. *The Guide to Health Informatics* (2nd Edition). Arnold, London.

Coiera, E., 2004. *Four rules for the reinvention of healthcare.* BMJ 328: 1197–1200.

Costello, P., 2002. *Intergeneration report 2002–3 budget paper No5.* Commonwealth of Australia, 2002. Available from: http://www.budget.gov.au/2002-03/bp5/download01_BP5Prelim.pdf [Accessed 19 February 2008].

Department of Health, 2007. *Commissioning framework for health and well-being.* Published for consultation 6 March. Department of Health. Available from: http://www.dh.gov.uk/en/Publicationsandstatistics/Publications/PublicationsPolicyAndGuidance/DH_072604 [Accessed 19 February 2008].

Duff, E., 2006. *Conference report. Resourcing global health.* Conference of the Global Network of WHO Collaborating Centres for Nursing and Midwifery Development, Glasgow, Scotland, June 2006. Midwifery Volume 22, Issue 3, September 2006, Pages 200–203.

Express Healthcare Management, 2001. *Robotic surgery defies distance barrier.* Available from: http://www.expresshealthcaremgmt.com/20011015/medtech2.htm [Accessed 19 February 2008].

George, M., 2001. *It could be you. A report on the chances of becoming a carer. Summary.* Carers UK, 2001. Available from: http://www.carersuk.org/Policyandpractice/Research/Itcouldbeyousummary.pdf [Accessed 20 November 2007].

Georgeadis, A.C., Brennan, D.M., Barker, L.M., and Baron, C.R., 2004. *Telerehabilitation and its effect on story retelling by adults with neurogenic impairments.* Aphasiology 18: 639–652.

Ghanta B., 2005. *Clinical outsourcing – Aussies join the India bandwagon the next big boom.* India Daily. 6th May 2005. Available from: http://www.indiadaily.com/editorial/2564.asp [Accessed 20 November 2007].

Gillies, B., 2001. *Smart Support at Home: an Evaluation of Smart Technology in Dispersed Housing.* University of Dundee, UK.

Gingrich, N., Pavey, D., and Woodbury, A., 2006. *Saving Lives, Saving Money.* Alexis de Tocqueville Institution, Washington, DC.

Health-EU Portal. Available from: http://ec.europa.eu/health-eu/index_en.htm

Hensel, B.K., Demiris, G., and Courtney, K.L., 2006. *Defining obtrusiveness of home telehealth technologies: A conceptual framework.* Journal of the American Medical Informatics Association 13(4): 428–431.

Hill, A.J., Theodoros, D.G., Russell, T.G., Cahill, L.M., Ward, E.C., and Clark, K., 2006. *An Internet-based telerehabilitation system for the assessment of motor speech disorders: A pilot study.* American Journal of Speech Language Pathology 15: 1–12.

IBP Initiative, 2007. *Implementing Best Practices in Reproductive Health.* Available from: http://www. ibpinitiative.org [Accessed 20 November 2007].

IHTSDO. International Health Terminology Standards Development Organisation. Available from: http:// www.ihtsdo.org

Israel, E., Chinchilli, V., Ford, J., Boushey, H., Cherniack, R., Craig, T., Deykin, A., Fagan, J., Fahy, J., and Fish J., 2004. *Use of regularly scheduled albuterol treatment in asthma: genotype-stratified, randomised, placebo-controlled cross-over trial.* The Lancet 364(9444): 1505–1512.

Johnson, P., 2006. *What are the limits of self-care supported by telemonitoring?* Telecare 2006, 11th May 2006, Coventry, UK. Available from: http://www.telecare-events.co.uk/2006/presentations.htm#johnson [Accessed 20 November 2007].

Jones, T.M. and Mead, C.N., 2005. *The Architecture of sharing. An HL7 version 3 framework offers semantically interoperable healthcare information.* Healthcare Informatics 22(11):35–36, 38.

Marley, T. and Rector, A.L., 2006. *Use of an OWL meta-model to aid message development.* Current Perspectives in Healthcare Computing (2006), Conference Proceedings, Harrogate, UK, March 2006.

Mitchel, D., 2006. Gazing into the crystal ball. In M. Conrick (Ed.), *Health Informatics: Transforming Healthcare with Technology.* Thompson/Social Science Press, Melbourne.

Mohr, F.W., Falk, V., Diegeler, A., Walther, T., Gummert, J.F., Bucerius, J., Jacobs, S., and Autschbach, R., 2001. *Computer-enhanced "robotic" cardiac surgery: experience in 148 patients.* Journal of Thoracic and Cardiovascular Surgery 121(5): 842–853.

NASA, 2005. *Behind the scenes: NEEMO 7: NASA Extreme Environment Mission Operations expedition.* Available from: http://spaceflight.nasa.gov/shuttle/support/training/neemo/neemo7/ [Accessed 10 December 2007].

National PACS Team (NPFIT) and National Radiology Service Improvement Team, 2005. *PACS Benefits Realisation and Service Redesign Opportunities.* Available from: http://www.radiologyimprovement.nhs. uk/%5Cdocuments%5Ckey_documents%5CPACS_Realisation_Final_May_IRB.pdf [Accessed 10 December 2007].

National Patient Safety Agency, 2007. *Putting patient safety first.* London: National Patient Safety Agency. Available from: http://www.npsa.nhs.uk/patientsafety/ [Accessed 20 November 2007].

Perlino, C.M., 2006. *The Public Health Workforce Shortage: Left Unchecked, Will We Be Protected?* American Public Health Association: Issue Brief. Available from: http://www.apha.org/advocacy/ reports/reports/

Perlman, A.L. and Witthawaskul, W., 2002. *Real-time remote telefluoroscopic assessment of patients with dysphagia.* Dysphagia 17: 162–167.

Ramnarayan, P., Tomlinson, A., Kulkarni, G., Rao, A., Britto, J., 2004. *A Novel Diagnostic Aid (ISABEL): Development and Preliminary Evaluation of Clinical Performance.* Medinfo 2004, 11(2): 1091–1095.

Rosen, J. and Hannaford, B., 2006. *Doc at a distance.* Spectrum Online. Available from: http://www. spectrum.ieee.org/oct06/4667 [Accessed 10 December 2007].

Seelig, M., Roberts, D., Harwin, W., Otto, O., and Wolff, R., 2004. A haptic interface for linked immersive and desktop displays: Maintaining sufficient frame rate for haptic rendering. In *17th International Conference on Parallel and Distributed Computing Systems* (PDCS 2004), *San Francisco, CA*, The International Society for Computers and their Applications (ISCA), Cary, NC.

Swan, G.M. and Godden, D.J., 2005. *North of Scotland TeleEducation Project Audit and Evaluation Report to NHS Education in Scotland.* University of Aberdeen Cerntre for Rural Health Research and Policy. Available from: http://www.northofscotlandtele-education.scot.nhs.uk/pdfs/Final%20Copy%20-%209 %20May.pdf [Accessed 20 November 2007].

U.S. Food and Drug Association, 2005. *Intuitive Surgical da Vinci Surgical System and Endowrist Instruments.* 510(k) summary. Available from: http://www.fda.gov/cdrh/pdf5/K050369.pdf [Accessed 12 June 2006].

W3C, 2001. W3C Semantic Web Activity. Available from: http://www.w3.org/2001/sw/

Waltner, C., 2004. *Space program NEEMO 7 relies on Cisco MPLS technology.* The Wi-Fi Technology Forum. Available from: http://www.wi-fitechnology.com/printarticle1637.html [Accessed 10 December 2007].

Wanless, D., 2002. *Securing Our Future Health: Taking a Long-Term View: Final Report.* HM Treasury, London, UK, 2002. Available from: http://www.hm-treasury.gov.uk/Consultations_and_Legislation/ wanless/consult_wanless_final.cfm [Accessed 20 November 2007].

Wikipedia, 2007. *Definitions for telemedicine, telehealth and telecare.* Wikipedia. Available from: http://www. wikipedia.com [Accessed 15 February 2007].

World Economic Forum, 2007. *Working towards wellness: Accelerating the prevention of chronic disease.* Presented at the World Economic Forum Annual Meeting, Davos, 2007. Available from: http://www.oxha. org/knowledge/publications/working-towards-wellness-wef-report-2007.pdf/view [Accessed 20 November 2007].

Wroe, C., 2006. Is Semantic Web ready for healthcare? In A. Leger, A. Kulas, L. Nixon, and R. Meersman (Eds), *ESWC'06 Industry Forum.* Budva, Montenegro.

14

Supply Chain Management in the Retail Sector

Edgar E. Blanco and Chris Caplice

14.1 Introduction

Retail firms sell goods and services directly to consumers utilizing a wide variety of formats and channels. Retail is the one of the largest industry sectors in the United States, accounting for 30% of the gross domestic product (GDP) and employing almost 20% of all US workers (Standard & Poor's, 2007). While it is a very mature industry, it is highly competitive with technology adoption and supply chain management (SCM) innovations being two prime strategies for competitive advantage. In 2005, for example, the retail industry spent $2.8 billion on supply chain software applications alone (Krishnan and Vanchesan, 2006). Studies have estimated that retailers who invest heavily in information technology can improve their productivity (measured as sales per worker) by 9% to 33% (Doms *et al.*, 2004).

The largest firm in the world, the retailer Wal-Mart, sold over $340 billion in fiscal year 2007, a figure larger than the GDP of Sweden. It is widely accepted that a big part of Wal-Mart's success has been driven by its innovation in SCM coupled with its judicious use of technology. These innovations include highly efficient cross-docking operations at its distribution centers and its unique collaborative supplier management system. Both of these require substantial investment in the use of advanced technologies. Its cross-docking operations, in which inbound shipments from multiple origins are re-sorted onto outgoing vehicles to multiple destinations with no storage of material, utilize exceptionally sophisticated optimization software developed specifically to handle its huge order volume. Its supplier management, on the other hand, relies on an ICT supplier platform called Retail Link that provides virtually all of its vendors with real-time point of sale and forecast data at the most granular level.

ICT Futures: Delivering Pervasive, Real-time and Secure Services
Edited by Paul Warren, John Davies and David Brown
© 2008 John Wiley & Sons, Ltd

Opportunities for the use of technology in retail are not limited to just the largest firms, however. Nordstrom Inc., for example, recently underwent a multi-year 'technology overhaul' investing in, among other things, a complete warehouse management system, a 'perpetual inventory' system, replenishment and markdown optimization software, a new point of sale platform, customer relationship management software, and a complete performance management dashboard application. The results from this ICT investment from 2001 to 2005 saw an increase in sales from $312 to $370 per square foot (*Retail Technology Quarterly*, 2006). Zara, a women's fashion retailer based in Spain, has combined its unique SCM approach with specialized ICT investments to outpace its rivals. Creating what they term a 'super-responsive' supply chain, Zara is able move products from design to the store rack in just 15 days, while the industry average is measured in months. Various technologies are deployed to assist this process, including customized handheld computers that allow retail store personnel to collaborate and communicate quickly to the marketing and design teams (Ferdows *et al.*, 2004). In these and many other cases, new technologies are utilized within the SCM function to drive significant cost savings, higher revenues, or better margins.

The objective of this chapter is to introduce the reader to SCM in the retail sector and to provide a framework for how technology is used currently and, potentially, will be used in the future. The remainder of the chapter is organized as follows. Section 14.2 provides a short overview of SCM in general. Section 14.3 follows with a description of the retail industry and the fundamental differences between retailers that impact both their ICT and supply chain strategies. Section 14.4 details how technology is utilized within the retail sector, while Section 14.5 summarizes the chapter.

14.2 Overview of SCM

SCM has many definitions. The Council of Supply Chain Management Professionals, the largest professional organization in this area, defines SCM as:

> ... the planning and management of all activities involved in sourcing and procurement, conversion, and all logistics management activities. ... In essence, supply chain management integrates supply and demand management within and across companies. (Council of Supply Chain Management Professionals, 2007)

The Supply Chain Council, another professional organization, provides a very succinct and operational definition stating that SCM is comprised of five distinct business processes: plan, source, make, deliver, and return (Supply Chain Council).[1]

The *plan* process focuses on demand and supply planning, design and configuration of the network, and alignment to the firm's overall financial goals. The *source* process encompasses all procurement and sourcing decisions including raw materials and semi-finished components up to the point of manufacturing by the firm. The *make* process involves production scheduling, materials management, and creation of all products, whether make-to-stock, make-to-order, or engineer-to-order. The *deliver* process covers the distribution of the product from point of storage or manufacture to consumption including order, transportation, inventory, and warehouse management. Finally, the *return* process contains tasks for handling the reverse flow of defective,

---

[1] http://www.supply-chain.org/cs/root/scor_tools_resources/scor_model/scor_model.

obsolete, or otherwise unusable product back to the firm or to final disposal. These five processes make up the Supply Chain Operations Reference (SCOR) model.

For retail operations, plan, source, and deliver are the most important SCOR model processes, while return is relevant in some cases, as well. However, as the next section explains, there are also some additional, retail specific processes that are critical to successful operations.

14.3 Overview of Retail Supply Chain

This section provides a short overview of retail operations and develops a framework that shows how various technologies may fit in to the overall process.

14.3.1 Key Processes and Flows

There are three core processes in the retail supply chain:

1. Merchandising – in charge of buying and selling goods. Includes decisions on product assortment, supplier selection, pricing, promotions, inventory allocation, and markdown. This is included in the plan process in the SCOR model.
2. Store operations – manages the point of contact with the customer. Includes decisions like store design, labor training, and staffing.
3. Distribution – the physical movement of products from suppliers to stores. This covers both the source, deliver, and return processes in the SCOR model.

Figure 14.1 shows how the three main retail processes interact with consumer and suppliers in a retail operation.

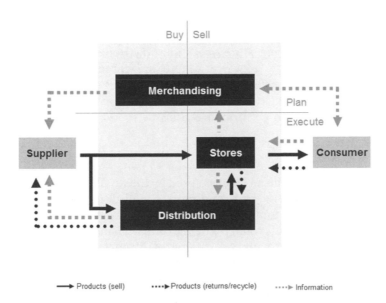

Figure 14.1   Key retail processes

Retail processes are usually divided into 'buying' and 'selling' activities. The merchandising functions span across both of these dimensions. Store operations deal primarily with selling activities. Distribution spans both of these activities. In addition, retail processes are usually divided into 'planning' and 'execution' activities. In the context of a retailer, planning is defined as the management of the sales and inventory to maximize profitability. Thus, merchandising encompasses the majority of the planning activities in a retailer, while store operations and distribution are considered execution activities.

There are three flows highlighted in Figure 14.1. The purchased products flow from the supplier to the store through the distribution network. Information is gathered from consumers directly or through the analysis of store operations to feed merchandising activities. Merchandising processes use this information to enhance buying and selling decisions. Notice that there is a two-way communication channel between merchandising and consumers: merchandising processes also provide consumers with information on product assortment (e.g. catalogs), seasonal calendars, prices, and promotions. Distribution processes use the store sales information to maintain the flow of goods according to the merchandise plan. The final flow is the reverse flow of goods. This could be returned merchandise (usually a small fraction of the total flow) or, increasingly, used products for recycling activities coordinated by the retailer on behalf of the suppliers (e.g. printer cartridges).

14.3.2 Fundamental Differences Between Retailers

There are thousands of retailers across the globe and they take many different forms. Retailers range from the behemoths like discounters Wal-Mart with almost 7000 stores to niche retailers with a single store. They also differ based on what and how they sell. Categories range from general merchandisers (such as Wal-Mart, Sears Roebuck, Federated, Target, etc.) that sell a wide array of products to specialty stores that focus on only one type of product (such as shoes, sporting goods, books, music, electronics, etc.). As the number of stores in a retailer's network increases, the complexity of the retail processes (merchandising, store operations, distribution) also increases impacting supply chain complexity. However, there are three fundamental differences between retailers that have as much impact on the retail supply chain as size; these are the channel, the product assortment between innovative and functional products, and the retailer's operational performance goals.

14.3.2.1 Retail Channels

The retail channel refers to the mechanism used by the retailer to service its customers. There are two main channels: 'brick-and-mortar' and direct. The traditional 'brick-and-mortar' store is the predominant retail channel: customers go to a physical store to browse and purchase the products. In the direct channel, customers review the retailer product selection, usually via catalog or electronically, and the product is delivered to the designated customer location.

The selection of retail channels has a significant influence on all retail supply chain processes. For example, for pure electronic retailers such as Amazon.com, store operations are replaced by Web-based software processes that can be easily adjusted based on supply and demand information. In fact, this is what allows some manufacturers, such as Dell, to enter and successfully compete in the retail space. Additionally, these direct retailers typically utilize sophisticated warehousing and inventory management techniques to enable efficient small order quantity fulfillment. Most retailers have a multi-channel strategy, but with the growth of the Internet, there are an increasing number of retailers that do not own brick-and-mortar operations but only sell via the Web like Amazon.com or Zappos.com.

14.3.2.2 Functional vs. Innovative Product Assortment

According to Fisher (1997), there are two main types of products that significantly affect supply chain strategies: functional and innovative. Functional products are those for which demand is predictable, with longer life cycles and lower margins. Functional products tend to satisfy basic needs and are usually manufactured by multiple suppliers. Innovative products have distinctive features (e.g. fashion or technology) that attempt to make them 'unique' for consumers. For this reason, these products tend to have highly unpredictable demand (will the product be a hit?) and very short life cycles.

As one would expect, the type of supply chain needed to support each type of product is very different. For functional products, the focus of the supply chain is usually cost efficiency. This includes tight control on inventory, negotiating volume discounts from suppliers, and streamlining distribution processes. For innovative products, the focus of the supply chain is responsiveness to customers. Therefore, maximum availability is the key to maximize profitability, and managerial focus is on maintaining very low stock-out rates and short lead times between suppliers and stores.

Retailers, of course, usually carry an assortment of both functional and innovative products. However, the mix or proportion of innovative to functional products carried impacts the retailer's supply chain operations. A retailer like Wal-Mart is a good example of carrying mainly an assortment of functional products and is therefore well known as having a highly cost-efficient supply chain. Its adoption of radio frequency identification (RFID) was primarily driven as a way to further reduce its costs of operations. Best-Buy or Zara, on the other hand, are good examples of retailers carrying mainly an assortment of innovative products. In fact, the very best retailers recognize the need to support different types of products and actually design and manage a portfolio of supply chains within their company. Technology plays a critical role in not only managing these operations, but also in the selection of which supply chain a product should be processed through.

14.3.2.3 Operational Performance Focus

But retail channels and product assortments cannot explain all of the differences between retailers. For example, convenience stores, such as 7-11, will carry grocery items as do pharmacies, such as CVS, and larger mass merchandisers, such as Target or Wal-Mart. The third important dimension affecting retail operations is operational performance focus.

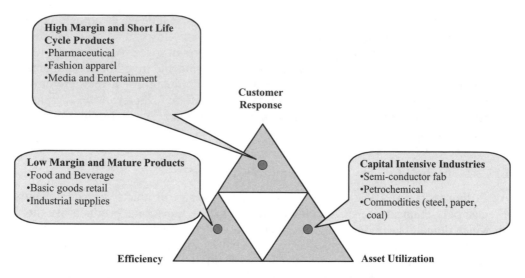

Figure 14.2   The operational objective triangle from the MIT SC2020 project

As part of the MIT Center for Transportation and Logistics Supply Chain 2020 (SC 2002) project, three main operational objectives have been identified: efficiency, customer response, and asset utilization (Lapide, 2006). These are the ways that a company measures itself and directs its resources. A firm that concentrates on efficiency utilizes inward facing productivity metrics such as supply chain costs, labor productivity, etc. Firms that focus on customer responsiveness are outward or customer facing and will use metrics such as order cycle times, quality, perfect order fulfillment, etc. Finally, firms focused on asset utilization will look to maximize the use of their capital intensive facilities. While all retailers are concerned with each of these three operational objectives, it is rarely equal between the three. In fact, the project argued strongly that in order to be competitive, a firm needs to concentrate its energy into one or two of these areas. Figure 14.2, shows the operational objective triangle used in the study.

The specific point where a company lies is determined by categorizing the key performance metrics used by senior SCM and then by calculating the proportion that falls into these three categories. For example, a firm with all efficiency based metrics will have a point in the bottom right corner of the triangle. Figure 14.3, illustrates where a number of companies fall.

While retailers will almost always carry an assortment of both functional and innovative products, the mix or proportion of one type over the other makes a difference in how a retailer will (1) set its operational performance goals (efficiency, customer responsiveness, or asset utilization), (2) design and operate its supply chain, and (3) implement information technology in support of its supply chain strategies[1].

---

[1] In practice, a retailer will establish multiple supply chains within its own operations to handle different product/customer/store combinations in the best manner.

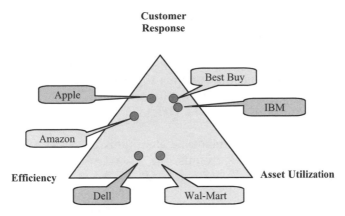

Figure 14.3   Sampling of company position on the operational objective triangle

While there might be a continuum of mixes and strategies, we can characterize the two extreme types of retailers as being efficiency-centric or customer-centric. We know of no retailers that are asset utilization-centric. Of course all retailers care about their customers and strive to be efficient; at issue is which of these two competing objectives trumps the other. This dichotomy helps highlight the different uses of and objectives for various technologies within retail operations.

14.4 Technology in the Retail Supply Chain

Retailers have been on the leading edge of SCM for several decades. They have been on the forefront of most of the SCM innovations including supplier collaboration, cross docking, vendor managed inventory, etc. This is primarily due to the fact that supply chain is one of their core competences. Retailers in general must source products built by others and distribute them to their various outlets. Underlying all of these operational innovations (Hammer, 2004) is the judicious use of technology and process design.

Successful technology innovations need to be aligned to the unique retailer characteristics and strategy. Their relative importance as well as the value of these technologies needs to be analyzed in the context of the retail factors presented in the previous section. The next two sections discuss different supply chain related practices and technology innovations in merchandising, store operations, and distribution for, first, the efficiency-centric retailer, and then the customer-centric based retailer.

14.4.1 Technology Focus for Efficiency-centric Retailers

The overriding objective for any technology adoption for efficiency-centric retailers is to lower the cost of operations. This can be achieved through improved execution with better merchandising, more streamlined distribution practices, and efficient store operations.

For many years, merchandise activities have been considered an 'art'. The ability to intuit customer preferences in fashion, product assortment, and pricing has always been managed by merchants with years of experience. Leveraging retailer investments on core information technology and point of sale data, science-based retail has gained momentum in the industry and will continue to grow in the future. Efficiency-centric retailers perform these tasks strategically and centrally. The use of technology will manifest itself in several ways:

- advanced forecasting and merchandising algorithms;
- promotion and price setting algorithms;
- assortment and allocation software.

Retailers with the capability to extract relevant consumer information from billions of transactions will maintain a competitive edge and will be able to streamline their operations. The first generation of applications for assortment optimization, forecasting, space optimization, regular, and promotional price optimization applications are currently available from several software vendors. However, as retailers have found out from deploying these solutions, leveraging the value of these applications requires additional ICT resources as well as ongoing training and support. The power and complexity of these applications as well as their hardware requirements will keep increasing. Software as a service models may help overcome some of these challenges as long as they can be properly integrated with the core business ICT infrastructure.

In-store operations for efficiency-centric retailers will utilize technology to improve the overall productivity of the labor and retail space. These include the following:

- Labor management and planning systems – Planning and allocation of staff have a significant impact on store performance.
- Enhanced point of sale systems – The actual sales at each store are captured and made available to vendors to help them improve their forecasting as well as their inventory replenishment activities.
- Promotion execution – A significant percentage of sales are lost because of poor promotion execution at the stores. Technology, like RFID, that helps track and manage the in-store promotion activities is being widely adopted.
- Enhanced in-store inventory visibility – The ability to track down products in the shelves and in the backroom allows for better deployment of merchandise throughout the supply chain and reduces stock outs. As the cost and standardization of RFID technology grows, some efficiency-driven retailers may be able to justify investment on inventory visibility.

However, efficiency-centric retailers will tend to utilize technology most frequently during the distribution activities. There are two basic kinds of distribution decisions: strategic and operational. Strategic decisions are usually made on an annual or longer basis. They involve extensive resources and expertise. Oftentimes, the retailer will require the use of outside consultancies or specialists to provide support for these decisions.

They include:

- Location analysis – Sophisticated mixed integer linear programming optimization and simulation software is used to determine where the various warehouses and distribution centers should be located. Store location decisions are not usually within the scope because of the nature of the retail business.
- Sourcing – Supplier selection is usually constrained by merchandise decisions, but understanding the impact of the multiple alternatives is usually part of the distribution.
- Distribution – This includes, for example, flow segmentation strategy for how specific products should move through the network, inventory deployment to determine how much product to stock at different echelons, and fleet planning to optimize the efficiency of private or for-hiring trucking assets.

Operational decisions are made on a regular basis ranging from daily to monthly. There are numerous software packages that automate these set of decisions. They include:

- shipment and pack size optimization;
- enhanced replenishment systems;
- dynamic product flow;
- shipment mode/carrier selection.

Besides the technologies that support strategic and operational distribution planning, the biggest technology investments are on distribution flexibility and visibility. Because of the nature of retail products, there is a significant time lag between merchandise decisions and store deployment: it is not uncommon in the fashion business for these decisions to be 18 months or more apart. Collaboration between distribution and merchandising activities is required to reduce this time. This collaboration is usually enabled by increased visibility of the products throughout the supply chain as well as flexibility of product movement. Examples include the following:

- RFID tags – For efficiency-centric retailers, the primary reason for RFID tagging is to reduce waste in the distribution process. The idea is to make a better bar code and to avoid misplacement of stock whether in different distribution centers, the backroom of the store, or on a truck. Wal-Mart, for example, proclaimed that they would generate an additional $3.6 billion in annual sales by stock-out reduction from implementing RFID tags (Duvall, 2007). For efficiency-centric retailers, tagging at the carton or pallet level provides sufficient benefit, while tagging at the product level is not as worthwhile.
- Advanced routing software – The complex process of dynamically selecting which order will be sourced from which distribution center and routed ultimately to the store requires very sophisticated software.
- Supplier collaboration and visibility tools – It has been estimated that Wal-Mart invested approximately $4 billion in its famous Retail Link (Chiles and Dau, 2005) that allows supplier visibility and collaboration. Increasingly, efficiently driven retailers will invest in similar technologies leveraging advances on Web-enabled platforms.

14.4.2 Technology Focus for Customer-centric Retailers

Customer-centric retailers will primarily use technology to enhance customer experience and improve service. Technology will be used to minimize the gap between their supply and the customer demand and to react quickly to customer behavior and preferences. Technology is allowing many of the merchandising decisions that were once made centrally and far in advance to be made at the local store level at the tactical or operational timescale.

Customer-centric retailers will utilize technology to tailor the retail experience to every individual customer. This includes customizing the product to the unique customer needs (also known as mass customization) as well as maintaining customer-level pricing and promotional activity. Merchandising activities will need to work even closer with suppliers, distribution, and store operations to achieve this level of customer intimacy. More and more of the merchandise activities will be done during 'retail execution' as opposed to 'retail planning'. Flexibility and visibility of the supply chain supported by a rich set of interfaces with stores and customers (i.e. Web, Internet, and mobile devices) will be a key success factor. In addition, the amount of data that will be gathered from all these sources will keep growing, increasing the complexity and the need for scientific retailing. For merchandising activities, the use of technology will manifest itself in several ways:

- higher resolution customer segmentation;
- increased use of 'loyalty' programs;
- versatile information systems will be able to manage higher granularity of customer buying patterns, and
- increased emphasis on multi-channel retailing to provide enhanced customer shopping experience.

In-store operations for customer-centric retailers will utilize technology to tighten the linkage between operations at the store and merchandising functions. Customer-centric retailing is the major trend affecting technology in-store operations. On one hand, store formats (i.e. size and layout) will start to vary significantly to accommodate local preferences. This increases pressure for store managers to be involved with merchandise decisions traditionally managed at the headquarters (e.g. space, assortment, and pricing). Given the execution focus of store managers, simplified interfaces and more mobile devices will be needed on the floor to support these new responsibilities. Increased collaboration between store managers and corporate merchandisers will also be required to maintain alignment with corporate strategy. Zara is already a pioneer on this area. Store managers have the ability to constantly influence their mid-season product assortment based on their own store performance, while the merchandising team at headquarters can electronically distribute specialized layout recommendations for each store in real time, e.g. see Chu (2005) or Caro and Gallien (2007).

Another effect of customer-centric retailing has to do with store equipment for capturing real-time information of customer buying patterns and behavior. For example, retail stores now have video cameras that not only serve security purposes. Advanced machine vision algorithms are able to track customer flow within the store.

This information is then used to improve store design and product placement. Another example is RFID technology that allows tracking of individual items throughout the store. This not only helps improve store operations (e.g. finding misplaced items) but also provides real-time information on shopping patterns (e.g. which two items were placed in the basket at any point in time).

Store technology investments are also focused on value-add services as well as enhancing the retail experience. By combining product identification technology (RFID or bar code) and a 'smart shopping cart', an array of services can be provided to the customer: price-checking, automatic coupon generation, shopping list management, or multimedia presentations shown during actual shopping. The MIT-AgeLab,[2] in collaboration with Procter & Gamble, has already developed a Smart Personal Advisor that uses the consumer's personal diet information to provide guidance at the point of decision – in the grocery aisle.

As a final note, the main barriers for store technology deployment in major retailers are inertia and scale: with thousands of stores and ten of thousands of store employees, deployment of any technology investment is usually a high risk, time-consuming, and capital intensive project. Thus, retailers are very risk adverse on adopting state of the art technology that affects all of their retail outlets.

For distribution activities, customer-centric retailers will utilize technology to minimize the gap between supply and demand. Most of the technologies described for efficiency-centric retailers are also applicable for customer-centric ones. The main difference is that the motivation on visibility and flexibility is to enhance customer experience as opposed to minimize cost. So for example, while enhanced inventory visibility through RFID may be used in Wal-Mart to collaborate with suppliers on finding efficiency, for Zara, this visibility will be used to develop new designs and increase assortment at their stores. This difference in focus makes the business case for technology investments a value or top-line driven proposition as opposed to a cost based decision.

14.5 Summary

Throughout this chapter, we have introduced the unique characteristics of the retail supply chain. We have also illustrated how the strategic focus impacts not only supply chain operations but also technology investments. The discussion of efficiency-centric and customer-centric retailers demonstrates the extremes of what is actually a continuum of retailer strategies. The retailers' investment in specific technologies must mirror and support their overall strategy and operational focus.

References

Caro, F. and Gallien, J., 2007. Inventory Management of a Fast-Fashion Retail Network, submitted to *Operations Research*.

---

[2] http://web.mit.edu/agelab.

Chiles, C.R. and Dau, M.T., 2005. An Analysis of Current Supply Chain Best Practices in the Retail Industry with Case Studies of Wal-Mart and Amazon.com. Massachusetts Institute of Technology, Master of Engineering in Logistics Program, Unpublished Masters Thesis.

Chu, P., 2005. Excellence in European Apparel Supply Chains. MIT-Zaragoza Program, Unpublished Masters Thesis.

Council of Supply Chain Management Professionals, 2007. http://cscmp.org/AboutCSCMP/Definitions/Definitions.asp.

Doms, M., Jarmin, R., and Klimek, S., 2004. Information Technology Investment and Firm Performance in US Retail Trade. *Economics of Innovation and New Technology*, 1 October 2004, p. 609.

Duvall, M., 2007. Wal-Mart Faltering RFID Initiative, *Baseline*, 2 October 2007.

Ferdows, K., Lewis, M., and Machuca, J., 2004. Rapid-Fire Fulfillment. *Harvard Business Review*. November 2004, vol. 82 issue 11, pp. 104–110.

Fisher, M., 1997. What is the Right Supply Chain for your Product? *Harvard Business Review*, March–April 1997, vol. 75 issue 2, pp. 105–116.

Hammer, M., 2004. Deep Change. *Harvard Business Review*, April 2004, vol. 82 issue 4, pp. 84–93.

Krishnan, S. and Vanchesan, A., 2006. World Retail Application software Analysis by Sector. *Frost & Sullivan Reports*, May 2006.

Lapide, L., 2006. The Essence of Excellence. *Supply Chain Management Review*, April 2006, pp. 18–24.

Retail Technology Quarterly, 2006. Wall Street's Darling. *Chain Store Age*, May 2006.

Standard & Poor's, 2007. *Industry Surveys Retailing: General*. 17 May 2007, page 9.

15

Technology Innovation in Global Financial Markets

Lesley Gavin

Global financial markets are dominated by international companies operating in a data driven environment, and represent one of the largest challenges for ICT vendors. After setting the background to this activity, this chapter reviews areas of technology that impact on this sector, discusses some special challenges and looks at the likely long term evolution as automation becomes widespread. The final section reviews the changes taking place that will shape the next decade of the market.

15.1 Introduction to the Financial Services Sector

The financial services sector has three distinct businesses: global financial markets (GFMs), essentially investment banking, including hedge funds and small investment houses; corporate banking, including debt markets, mergers and acquisitions; and retail banking, essentially high street banking, lending and insurance services. The sector as a whole has long been an intensive user of technology. Stock market 'ticker' information brought updated Wall Street information to the Main Streets area of New York as early as 1867, and robotic or automated tellers have been around since a Long Island branch of Chemical Bank installed the first in 1969 (Kauffman and Webber, 2002). Within the sector, GFMs have the highest ICT expenditure and are often an early adopter of technologies. It is for this reason that this chapter will focus on GFM.

The GFM sector has a number of interesting characteristics. It is a global market-place, dominated by international companies operating in what is essentially a data driven environment. In this respect, it is ideally placed to take advantage, and quickly benefit, from ICT technologies. It regards itself as a highly specialised sector, with individual needs, hence the historic proliferation of in-house ICT development. Key

ICT Futures: Delivering Pervasive, Real-time and Secure Services
Edited by Paul Warren, John Davies and David Brown
© 2008 John Wiley & Sons, Ltd

technologies used range from order processing systems, including clearing and settlement, vast databases, complex back office systems to trading optimisation and risk management systems (Freedman, 2006). The sector handles extremely large volumes of real-time data processing. Increasing automation has led to increased investment in ICT technologies. Two examples of automation are straight-through processing (STP) aimed at reducing costs, and algorithmic trading systems which aid new product development. Both are discussed in more detail in Section 15.3.2. The highly global, knowledge based nature of this sector also makes it an interesting study as regards the implementation of emerging distance working and collaboration technologies. This is discussed in more detail in Section 15.3.3.

15.2 Key Technologies for the Sector

Figure 15.1 illustrates the four technologies that are driving change and creating opportunities in this sector: data collection, data storage, data processing and data display. As each technology progresses, it opens new possibilities for improving existing systems or creating new applications. Technology choices shape the innovation possibilities. Technology evolution is taking place within each of the four areas of the framework – but when the four areas of technology are considered together, there is a very definite change taking place which is likely to transform the way that business is done in this sector.

15.2.1 Collection of Data

Technologies that collect data, that is, those that record voice, text, video, etc., are becoming more ubiquitous. In our everyday lives, we are seeing an increase in closed circuit television (CCTV), home security systems with wireless sensors, access control systems. We can track parcels as they travel around the world, and our cars now give us real-time dynamic travel directions. In our work environment, it is now possible to record all texts we write, phone calls we make and conversations we have. Sensors are important in convergence because in order to deliver information to us in the right

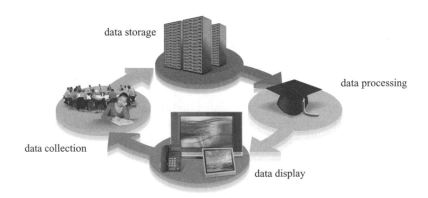

Figure 15.1   Four key technology fields

format, the converged system has to know where we are, and exactly how we would like to receive the information. The barriers to the deployment of sensor technology are very real. Privacy issues are a major concern. Some cultures are more accepting of this than others, but in general, we are uneasy about everything we do and say, where we are and when, all being recorded. But in this sector, regulation and compliance mean that employees are already heavily monitored.

15.2.2 Storage of Data

Data storage is already important for this sector. The sector is data intensive. Information has to be kept securely. It also needs to be accessible for audit to meet with regulatory and compliance rules. The proliferation of sensor technologies will lead to an increased volume of data that will put further pressure on the provision of storage facilities. The sector has very stringent business continuity demands, so all data need to be kept in multiple locations and accessible from multiple locations. Virtualisation technologies that allow data to be securely stored in different locations will be increasingly in demand. Technologies that reduce the cost of storage (for example, optical storage solutions) will be in demand. Storage also brings a set of environmental challenges, and as companies become more environmentally sensitive, businesses will need to adopt approaches that minimise the consumption of carbon. Storage is also increasing in local devices (including mobile phones), and so technologies that manage content wherever it is in the organisation are likely to be deployed in the longer term.

15.2.3 Processing of Data

The GFM sector demands real-time processing. Milliseconds count.

Currently, data processing is primarily focussed on transactional processing and varying types of expert systems. Expert systems can analyse a particular company's finances, drawing on data describing profitability, collating ratios and correlating over time. They can also be used for market analysis. New and emerging processing technologies aid not only quantitative processing, but also qualitative processing that involves understanding of the meaning of the data. This could include recognition of relationships between events, and will require the adoption of semantic technologies that seek to understand the underlying meaning of information. Semantic technologies are undoubtedly going to have a major impact in any knowledge based industry.

New technologies will be able to extract information from any available source, e. g. standard databases, but also the web including edited articles and also blogs, etc. This paves the way for 'sentiment analysis', a concept where information on emerging trends becomes readily available. An example of this could be tracking levels of consumer grievances against a particular company through the monitoring of blogs and consumer forums.

15.2.4 Display Technologies

Once all this processing, relating and understanding has happened, the information, or knowledge, needs to be delivered back to the individual. Display technologies are evolving rapidly. Flat screen displays are now everywhere, and in the next decade, the emergence of 'digital paper' is likely to provide many more display options. Mobile devices are now very widely used in many industries, and will be increasingly used within this sector.

Technologies are being developed that understand the context of the individual or group, where they are, who they are with and how best to communicate with them at this moment in time. The means of communication could be large displays, small devices or wearable appliances depending on the nature of the information and of the environment of the individual. Sometimes, a text message could suffice; in others, it might be that large scale images are required. The kind of display used might depend on the security settings required for a given situation, taking account, for example, of whether individuals are in the privacy of their own home and can openly display any information, or they are on a crowded train and would only want to display appropriate information. The way information is delivered will also depend on the nature of the response required: is an immediate response expected, perhaps in an emergency situation, or is it more a case of making individuals aware of a situation that they can look into more closely later. Context awareness and mobility are two key areas of technology that are likely to impact on this sector.

15.3 Key Issues for GFM

The GFM is a highly data-centric sector, with data being collected from multiple sources including exchanges, news agencies, in-house and out-of-house analysis, real-time and otherwise. It is a global workplace, indicating a high degree of mobility, the importance of time zones and the importance of cross-cultural understanding.

This section explores three issues that are fundamental to the industry when thinking about the way businesses are likely to invest in technology.

1. Regulation is a major shaper of activity in this space.
2. Automation as this underpins the drive to manage costs, as well as allowing companies to grow faster.
3. Changing working environments as the future brings some real challenges to existing methods of working.

15.3.1 Regulation

Regulation and political developments are both imposing increased operating overheads, and changing the market structure and competitive dynamics of the industry. European harmonisation, for example, covers not only the more visible Basel II and International Finance Reporting Standards (IFRS), but also a whole raft of other areas such as those related to securities participants market conduct, cross border securities settlement, money laundering and electronic settlement. The convergence of exchanges and market infrastructure will also continue (Barrett and Scott, 2000). GFM organisations are global, operating in many countries and multiple regulatory areas, with each regulatory body around the globe setting its own requirements.

The highly regulated nature of this sector also, rather interestingly, removes a prime barrier for the technology innovator: Regulation means that people are both familiar and accepting of all voice and data transactions being recorded and monitored. In other sectors, privacy of information laws, and general social backlash and fear of 'big brother' prohibit the collection of such data. As our earlier framework shows, data collection will be key to being able to identify areas where new technology can be effective. This sector offers a great test bed for emerging applications.

15.3.2 Automation

As with most businesses, there is an ever increasing pressure to cut costs and to increase revenue. In this data-centric environment, there has been ample opportunity over the last decade to cut costs by automating paper driven activities. STP has eliminated many manual tasks by automating order execution and clearance.

Processing technologies are an extremely important area for this sector, and the sector has a relatively long history of investing in these technologies. Artificial intelligence (AI) has been commercially used by credit agencies for over 20 years. A further technology leap was taken in 1996 when the London Stock Exchange installed a package of AI software to examine share dealings to identify insider trading and other fraudulent activities (Gosling, 1998).

Then, in 1998, the advent of electronic trading systems and electronic exchanges opened the doors for end to end electronic trading (Barrett and Scott, 2000). This resulted in an explosion of 'algorithmic trading', or in other words, algorithms that attempt to predict market movements and subsequently trigger the automatic execution of buying and selling shares. AI had been used with varying degrees of success: the 1987 Wall Street Crash was in part blamed on a number of dealing rooms using software that issues instructions to sell when the price dropped, but didn't ever buy the shares back. In 1996, Switzerland's electronic stock exchange had to suspend trading several times because the software generating the buy and sell instructions was causing massive price fluctuations (Gosling, 1998).

Today, applications of AI in GFM use both neural networks and increasingly genetic algorithms. Genetic algorithms are useful because they are able to give a clear history of reasoning behind the decision making, which helps with regulation and compliance as well as with general understanding. Some algorithms that show market states (rather than make decisions) are available through information agencies such as Reuters or Bloomberg, but the most enticing algorithms are those developed by expert mathematicians working with the traders with the aim of a fully automated trading system that reacts expertly to changing market conditions. Most are developed in-house and are usually proprietary. These algorithms are now highly sophisticated instruments dealing with both high and low levels of risk. They allow new innovative products to be created. There is, however, a constant concern that the global nature of financial markets and the proliferation of such algorithms from numerous banks and investment houses mean that no one really knows how interlinked they all are, or exactly where the risk is.

To date, the inputs for algorithmic trading have been based on historical exchange and market based data. However, we all know that the past prediction of markets is no indication of the next step. This is where semantic technologies can help:

If we take the simplified example of trading the shares of one company (rather than an index), it would be preferable to have all the information about everything that has been happening in that company, not just historical trading information, but up-to-date real-time information on company strategy and operations. Even now, there is actually a lot of current information around about companies – in company reports, announcements, newsfeeds. There is even a prolific amount of the less formal company information in random blogs – information that perhaps gives a more immediate understanding of current customer experience. All this information could be viewed as in indicator of future performance and could be fed into new types of algorithms. The processing algorithms would of course have the mammoth task of understanding the relevance of the information received. Relevance can be related to context, which is related to levels of confidence in the source. The relevance of an event will also change depending on other related events or pieces of information. One complaint on one blog may be relatively meaningless, but thousands of complaints may be an indicator of something going wrong. On the other hand, if the company is going through a change period, perhaps a certain level of customer dissatisfaction is acceptable, indeed expected, and therefore a positive indicator of change.

The next level would of course be to aggregate this knowledge throughout a portfolio, perhaps compare the likely differences in the company's ability to react to micro and macro economic events, etc.

Taking a step back, if we look at this sector, we can identify tasks that are essentially process or transaction driven. These tasks are prime targets for being taken over by machines. The results of much faster, more dynamic processing should theoretically be less imperfect markets, and therefore less margin to be made by the banks. IT investment is expensive, and while it seems odd to spend so much on technology that will make it harder and harder to make money, competition dictates that it is necessary, as those who don't invest in this technology go out of business.

15.3.3 Changing Work Environment

So if machines are doing all the work, what will the bankers actually do? If you ask traders, they will freely admit that their job won't exist in the future (again you are less likely to hear similar comments from any other sector – probably more due to job protectionism rather than lack of recognition of the possibilities of technology). What they mean when they say this is that the processing part of their job won't exist. If we look again at the scenario above where semantic technologies are used to analyse companies, there is a very big question mark today over the validity and trust of the sources of information. Some sources can probably be trusted automatically, e.g. information direct from the Federal Reserve, but others need to be continually assessed. In some situations one individual's opinion could count more than others, the future bankers job will be to discern and make decisions on the importance and relevance of some of the information used as input sources to the algorithm. Essentially, the banker's job will be more about assessing levels of trust in information.

The activities that take place in this environment range from the highly collaborative to entirely process driven. Activities can be seen as process based, knowledge based and people based. The first lends itself easily to automation. There is a recognised understanding of how technology can make these activities faster and cheaper. Developments in processing time, network latency and AI will have a big impact on process based activities. New technologies are only just beginning to have an effect on knowledge based activities, and we have only just touched the surface of enhancing people based activities.

The nature of large corporations is also changing. The modern corporation needs flexibility; it needs to minimise fixed costs, changing these to variable costs as much as possible (Gosling, 1998). This is alongside the skills based required becoming increasingly specialised, and customers expecting increasingly personalised services. We can see trends in employment patterns that lean towards corporations having a relatively small core staff with say 'annual hours contracts' rather than the fixed 9 am–5 pm (or 7 am–7 pm in the case of this sector) and a vast increase in freelance workers brought in for particular specialist tasks.

The workplace means different things to different people, at different stages of their lives. Young people like their workplace to be a social place, where they meet and make new friends. As we grow older and have families, workplace can be viewed as a place of relative calm and order, while earning a living to support the family becomes increasingly important. Working mothers are an example of a group that sees the workplace as a place to be efficient, a place to do tasks quickly, and they appreciate being able to leave when they have finished. The face-time, long hours based environment of the traditional investment bank is highly geared to the profile of those they want to attract: the aggressive young male. Gartner[1] has continually listed employee retention in the bank's top 10 greatest concerns for a number of years, indicating that this environment does not retain staff as they get older and move into different phases in their lives. In fact, the recent surge of hedge funds has picked up on this. The work environment in a hedge fund is relatively short hours, more flexible working patterns, and all heavily supported by technology.

Technologies that support flexible working for knowledge workers, working in a global, culturally diverse, highly specialised work environment are likely to grow enormously over the next few years as institutions increasingly need to retain key specialist staff. However, the introduction of technology alone will only have a minor impact in the first instance. To gain the greatest economic impact from these technologies, they have to be integrated into the culture of the organisation, and cultures tend to take a long time to change.

15.4 Conclusion

Financial services are, and always have been, a particularly specialised area, hence the widespread use of in-house developers. But with rapid new technology development, and increasing pressure for these institutions to keep ahead of the curve, it is increasingly unfeasible for banks to be technology experts.

The exploration of the technology framework in Section 15.2 allows us to see how the financial services sector is a prime sector for the exploration of new technologies. In an earlier chapter, Anderson and Stoneman advocate an ICT innovation process

---

[1] http://www.gartner.com/

that involves user participation, and a development process that allows for iteration and adaptation. User participation in design and development is not a new process, it is the norm in industries such as consulting, building design and product design, but is a challenging one. The iterative design process has many advantages; not least, it is a way of integrating specialist knowledge into complex system development and reduces the risk of failure by constantly checking that the system will meet its requirements. It often has the additional benefit that as the parties discuss potential, other more innovative areas of application can be discovered. The art of innovation, that is, the application of technology to an area to create something new, has a much better chance of succeeding through partnership and collaboration.

References

Barrett, M and Scott, S (2000) 'The emergence of electronic trading in global financial markets: envisioning the role of futures exchanges in the next millenium.' In H Hansen, M Bichler and H Mahrer (eds.) *ECIS 2000: A Cyberspace Odyssey: Proceedings of the European Conference on Information Systems (8th), 3–5 July 2000, Vienna University of Economics and Business Administration, Vienna, Austria.* Vienna: Vienna University of Economics and Business Administration, pp. 717–722.

Freedman, RS (2006) Introduction to Financial Technology, Amsterdam, Elsevier.

Gosling, P (1998) Financial Services in the Digital Age, London, Bowerdean.

Herriot, P, Hirsch, W and Reilly, P (1998) Trust and Transition: Managing Today's Employment Relationship, Chichester, John Wiley & Sons, Ltd.

Kauffman, R and Webber, B (2002) Introduction to the Special Issue on Advances in Research on Information Technologies in the Financial Services Industry, USA, Journal of Organisational Computing & Electronic Commerce, Vol. 12.

16

Technology and Law: The Not So Odd Couple

Marta Poblet, Pompeu Casanovas and Richard Benjamins

16.1 The Impact of Technology in the Legal Domain

Practising law is an intensive knowledge task, and technology has also gained a solid footing in the area of knowledge management (Baldwin 2007). Everywhere in the legal domain, lawyers are knowledge workers with an ever-increasing demand for IT solutions. In some areas – notably, document management, case management, time recording systems, and legal information search – technology has already provided software systems and databases that have become standard in the legal profession, adapting to the needs of the online environment. As regards legal research, in many countries, the legal profession has online access to nearly every piece of legislation. Lawyers may also access a massive number of judicial decisions of courts either through public or fee-based legal information providers (i.e. Westlaw or LexisNexis). No doubt, technological innovation has changed the way lawyers prepare their cases and draft documents. But technology is also having a tremendous impact on how lawyers communicate and interact within online environments. Not only e-mail has been adopted as the standard way to communicate and exchange documents, but according to some reports, it has also become the repository of 60–70% of business critical data (Smallwood 2006). Besides, an increasing use of instant messaging (IM), Voice over IP, web meetings, corporate intra and extranets, blogs, wikis, etc. tends to fragment and disseminate legal information through multiple sources. This 'explosion of content', to quote Smallwood (2007, p. 1), inevitably generates new demands on how to search, manage, store, and retrieve legal information, be it formally contained in corporate documents or dispersed informally in those multiple sources. To fit in the extremely competitive global environment, law firms need to address carefully each of those issues. We review some of the most usual strategies in the section below.

ICT Futures: Delivering Pervasive, Real-time and Secure Services
Edited by Paul Warren, John Davies and David Brown
© 2008 John Wiley & Sons, Ltd

16.2 ICT in Law Firms: Recent Trends

A few years ago, very little information could be found in the legal domain on issues such as technology uses in law firms, ICT budgets, consumer satisfaction, etc. Today, annual technology surveys are part of the legal marketplace and cover a wide range of ICT topics:[1] computing technologies in law offices and courtrooms, online research, electronic data discovery (EDD), web-based and mobile communications, etc. While covering different segments of both national and international legal markets, some trends may be highlighted from these multiple sources. To complete the landscape, we include in this section the results of research focusing on the Spanish legal market (Consejo General de la Abogacia Española 2006).

16.2.1 ICT Budgets

According to the 2006 American Bar Association Tech Report (ABA 2006), 60% of law firms budget specifically for technology (Ikens 2007, p. 2). Ikens also observes that this figure is correlated with firm size, since only 35% of solo practitioners specifically budget for technology, compared to 90% for firms of 50 to 99 attorneys, and to 85% for firms with over 100 attorneys (2007, p. 2). In most firms, all partners make technology decisions (23%), followed by the managing partner (22%), a technology committee (15%) and the office manager (12%) (2007, p.2). As regards technology purchases, results from the 2007 ILTA survey show that small firms (under 200 attorneys) have 'higher implementation rates for case management, courtroom technology, docketing software, imaging/scanning/OCR, patch management, and records management software', while large law firms 'dominated in the implementation of OS upgrades, remote access technology, voicemail upgrades, workflow automation, and wireless devices' (ILTA 2007b, p. 2). Finally, data from the 12th AmLaw Tech survey, polling Chief Information Officers (CIOs) and IT directors from 200 largest American law firms show that those firms spend on average almost $33 K per lawyer (2007, p. 1).

16.2.2 Computing Technologies

Lawyers and legal advisers' use of computers is virtually universal. In Spain, use of computers is now at 99% (Ministry of Industry-Red.es 2004). The 2006 ABA Tech Report offers more details as regards the type of primary computer used at law offices (Figure 16.1).

16.2.3 Online Research

Legal research is also mostly conducted online. Ninety-three percent of Spanish lawyers and 96% of American attorneys already used electronic resources online in 2005

---

[1] Consider, for instance, the recent 2007 American Bar Association (ABA) Legal Technology Survey Report, the 2007 ILTA Technology Purchasing Survey, and the First ILTA Telecommunications Survey (ILTA 2007a). In nearly all cases, access to full reports is fee-based or restricted to members of legal associations.

	Number of lawyers at all locations					
	Total	**Solo**	**2–9**	**10–49**	**50–99**	**100 or more**
Desktop computer	70.5%	75.1%	78.7%	72%	67.5%	55.2%
Laptop computer	15.5%	19.9%	14%	13.7%	17.8%	13.8%
Laptop with docking station	13.4%	5.1%	6.9%	13.5%	14.2%	29.9%
PDA/Smartphone/ Blackberry	1%			2%		4%
Tablet PC	3%			7%	6%	4%
Other	2%		3%			4%
Total	100%	100%	100%	100%	100%	100%
Count	2332	534	649	453	169	522

Figure 16.1 Primary computer used at work (Reprinted with permission from the January 2007 edition of LJN's Legal Tech Newsletter © 2007 ALM Properties. All rights reserved.)

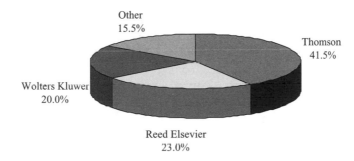

Figure 16.2 Relative market shares of major legal publishers in 2006

(CGAE 2006, Palmer 2006). These resources are either fee-based or free, but as Gelman puts it, 'lawyers have long demonstrated willingness to pay for up-to-the minute access to information in an easily accessible format' (Gelman 2004). According to Ikens, 'for the legal profession, the high use of a fee-based source could be attributed to the fee-based resource's authority and recognized brand name within the legal marketplace' (2007, p. 4). Although there are a number of legal information providers, Thomson (Westlaw), Reed Elsevier (Lexis-Nexis), and Wolters Kluwer, together known as the Big Three, control about 85% of the market for legal information (Figure 16.2).

It is important to note that research is not necessarily done by attorneys themselves. When surveyed, only 57% of them report that they personally conduct legal research (Palmer 2006). Especially in large law firms, attorneys are normally assisted by other attorneys (71%) or firm librarians (35%) in research tasks, while law office clerks and summer associates provide research assistance 53% of the time (Palmer 2006). Nevertheless, as Palmer notes, 'lawyers are no different from the rest of the population by selecting Google as the search engine they use most often' (2006, p. 2). Other surveys in Spain confirm the combination of legal databases and Google to perform legal queries (i.e. CGAE 2006).

16.2.4 EDD

EDD or e-discovery 'is the process of collecting, preserving, reviewing and producing electronically stored information (ESI) in response to a regulatory or legal investigation' (Murphy 2007, p. 1). In the USA, e-discovery has become a hot topic with the December 2006 amendments to the US Federal Rules of Civil Procedure, which governs procedures for civil suits in district courts. The rules introduce the notion of 'ESI', which may be interpreted as 'everything from program files and voice mail to e-mail, websites and instant messages' (Swartz 2007, p. 24), including malware and spyware. Notably, the new rules recognize ESI as distinct from paper documents, and lay out the ground rules for the EDD process from the outset of a law suit. While the impact of the rules is not yet evident in surveys, recent data show that EDD is gaining relevance in law practice. The 2007 ABA Legal Technology Survey reports that 16% of firms receive EDD requests three to 11 times a year, and another 13% two times a year or less. While 57% of attorneys have never received EDD requests on behalf of their clients, this average was 62% in 2006 and 73% in 2005; on the production side, only 26% of attorneys never make electronic discovery requests, versus 69% in 2006 (ABA 2007). Litigation support software is currently developing new tools to match these demands.

16.2.5 Mobile and Web-based Communications

Cell phones and wireless e-mail devices, such as BlackBerrys or personal digital assistants (PDAs) are part of the lawyer toolkit. Nevertheless, while more than 95% of lawyers regularly use cell phones, less than half use PDAs (Poblet *et al.* 2007), although these are more frequent in large law firms. These figures coincide with the 2006 ABA survey: while almost half (49%) of respondents have smartphones or BlackBerrys available to them, only 18% of solo practitioners – but 88% of large firm respondents – have them available (Ikens 2007, p. 3).

Web-based communications have also been adopted in the daily practice of law. According to data from the 200 AmLaw annual survey, web conference software ranks first, with 67% of the firms surveyed using it for administrative meetings, client meetings, in-house training programs, and communication with colleagues (Violino 2007). In contrast, 46% of firms report that IM is prohibited, due to internal policies to minimize the risk of inappropriate exchanges in terms of e-discovery.

Firms also tend to increase their use of extranets to communicate with clients, but other collaborative technologies such as intranets and wikis are less common. As regards legal blogs (or *blawgs*),[2] lawyers still do not use this tool: on the one hand, only 5% of 2007 ABA respondents have a blog, which is generally maintained by a single lawyer or a group of lawyers of the firm; on the other hand, over half of respon-

---

[2] See Conrad and Schilder (2007) for an opinion mining analysis of the legal blogosphere.

dents never read blogs for current awareness (22% less than once a month, 12% one to three times per month, and 12% once or more in a week). The use of Really Simple Syndication (RSS) feeds is even less frequent: 83% never use them, and only 5% one or more times a week. Finally, the use of podcasts is still minimal: 3% of respondents use podcasts for current awareness one or more times per week versus 80% of lawyers who never use podcasts.

16.2.6 XML Technologies

Currently, the adoption and development of standards for legal information, electronic court filing, court documents, transcripts, criminal justice intelligence systems, etc. has become the core activity of a number of initiatives and projects. To quote some examples, the nonprofit OASIS LegalXML (a subgroup within OASIS) was created in 1998 to develop 'open, nonproprietary technical standards for structuring legal documents and information using XML and related technologies' (OASIS LegalXML 2007). Technical committees within LegalXML work on areas such as court filing, e-contracts, e-notary, international justice, lawful intercept, legislative documents, and online dispute resolution (ODR). In Europe, the LEXML community defines itself as a 'European network searching for the automatic exchange of legal information' (Blanco and Martínez 2007, p. 30). Developing European standards includes learning lessons from previous national projects such as Norme in Rete (Italy), Metalex (the Netherlands), LexDania (Denmark), CHeXML (Switzerland), or eLaw (Austria).

16.3 E-justice and ODR

In very few years, ICT has pervaded all areas of the judicial business. Both e-justice and ODR have emerged as novel fields at the crossroads of law, conflict resolution, computer engineering, artificial intelligence (AI), and knowledge management. Usually, the term 'e-justice' applies to the specific developments of ICT in judicial environments (i.e. access to courts, case management systems, case-based reasoning, legal expert systems, e-mailing of legal notices, e-filing, e-judgments, e-signature, e-briefs, videoconferencing, etc.), while ODR can be defined in a broad and flexible sense to encompass not only disputes that originate from online transactions, but also off-line disputes handled online. Similarly, the 'online' component may be extended to include the use of electronic applications such as videoconferencing, mobile telephony, voice over IP, etc.

E-justice tends to focus on public administration reform as a way of thinking how ICT may contribute to organize judicial systems and their institutions as efficient public services. A felicitous example of how ICT may facilitate access to justice is the Internet service 'Money Claim Online' of UK courts, allowing both claimants and defendants to make online claims or respond online to claims for a fixed amount (i.e. an unpaid invoice).[3] Another recent example of ICT applied to better access and organize expert

---

[3] See how Money Claim Online works at https://www.moneyclaim.gov.uk/csmco2/index.jsp.

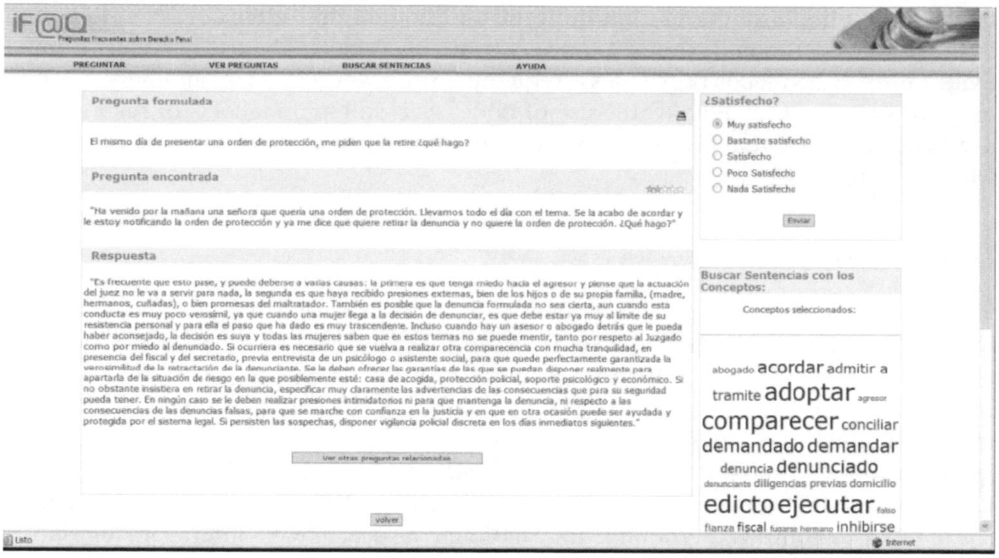

Figure 16.3   Iuriservice frequently asked question search system

judicial knowledge is Iuriservice, a web-based system intended to provide the Spanish judiciary with a tool to facilitate knowledge management in daily judicial practice (Figure 16.3).[4]

ODR systems, usually labeled as 'out-of-court' dispute resolution, seem to find their most successful achievements in the e-commerce area and other online transactions. At the present moment, several ODR providers are developing new paradigms for dispute resolution models on the Internet: notably, SquareTrade, Cybersettle, World Intellectual Property Organization (WIPO) domain name arbitration, or the Wikipedia). Despite their particularities, they all tend to make conflict management processes more efficient and speedier for external users and legal professionals alike. As Katsch and Wing put it, 'the new challenge is finding tools that can deliver trust, convenience, efficiency, and expertise for many different types of conflict' (Katsch and Wing 2006, p. 21) (Figure 16.4).

Another important trend in the domain of e-justice is the increasing number of software solutions adapted to litigation activities (both pre-trial and trial procedures). For instance, trial presentation software for lawyers at court or mediators in settlement conferences help to create paper-free presentation of evidence (Newman 2006). Litigation support software also helps to search, scan, tag, and file documents. Nevertheless,

---

[4] Iuriservice has been developed in the framework of the Semantic Knowledge Technologies Project, European Union Information Society Technologies Integrated Project, VI Framework Program (EU IST IP 2003-506826). Iuriservice is a web-based application that retrieves answers to questions raised by incoming judges in the Spanish judicial domain. Iuriservice provides these newly recruited judges with access to frequently asked questions through a natural language interface. The judge describes the problem at hand, and the application responds with a list of relevant question–answer pairs that offer solutions to the issue, together with a list of relevant judgments (for a general description of Iuriservice, see Casanovas 2007; see also Poblet *et al.* 2007).

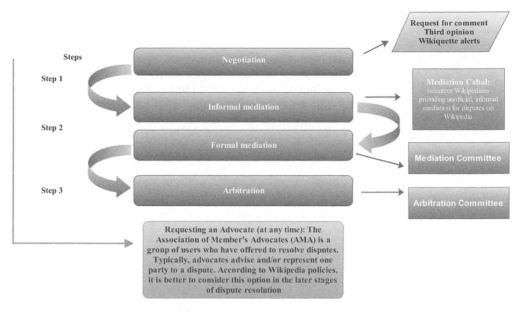

Figure 16.4 Wikipedia ODR system

	Number of lawyers at all locations					
	Total	**Solo**	**2–9**	**10–49**	**50–99**	**100 or more**
Yes	29.5%	6.8%	13.9%	37.4%	50%	67.7%
No	53.2%	90.7%	77.6%	32.2%	13.2%	2.4%
Don't know	17.3%	2.5%	8.5%	30.4%	36.8%	29.9%
Total	100%	100%	100%	100%	100%	100%
Count	1727	442	496	342	114	331

Figure 16.5 Litigation support software in law firms (Reprinted with permission from the January 2007 edition of LJN's Legal Tech Newsletter © 2007 ALM Properties. All rights reserved.)

and despite the multiple products available in the market, the figures are not impressive yet. In the USA, less than a third of respondents have litigation support software at their firm (Ikens 2007) (Figure 16.5).

16.4 The Special Case of E-mail Management (EMM) in Law Firms

Use of e-mail services has become ubiquitous in the legal profession. According to the 2007 ABA Legal Technology Survey Report, 97% of attorneys said they use e-mail for routine correspondence, and more than 70% reported they use e-mail for case status, memoranda, or briefs (ABA 2007). E-mail is also used by many attorneys for client billing, court filing, and marketing.

Email is pervasively the client communication standard for law firms and professional services firms, often in conjunction with collaborative extranets housing consolidated billing and matter status information [information about the case]. Email is frequently the de facto workflow vehicle for document revision, with versions sent for review to progress contracts, briefs and agreements to a state of acceptance and completion. In addition to the inherent confusion that can ensue with lax content process control, firms must operate in compliance with guidelines that address new business intake, from government mandates (USA PATRIOT Act, UK Financial Services and Markets Act, EU Privacy Laws, U.S. Safe Harbor) to due diligence in conflict of interest research (Kersey 2006).[5]

Law firms such as Freshfields Bruckhaus Deringer (2400 lawyers in 18 jurisdictions) have reported to send and receive more than 1.75 million e-mails in a typical month (Dineen 2006). No surprise, therefore, if it is becoming increasingly difficult for lawyers to deal with ever larger flows of e-mail communication. Considering its multiple uses, the tasks of locating, indexing, archiving, or retrieving e-mail will become harder for lawyers in the immediate future.

In addition, both inbound and outbound content compliance (OCC) are becoming critical requirements for any EMM system within law firms. OCC implies the detection and prevention of outbound content that violates policies of the organization and/or government regulations. Therefore, OCC deals with internal threats, as opposite to more traditional security solutions (firewall, anti-virus, anti-spam, etc.) dealing with external threats. Since the provision of legal services is one of the most scrutinized areas, lawyers need also to be aware that sending, receiving, and holding e-mail correspondence may involve the processing of personal data which must be dealt with in accordance with data protection legislation. According to the guidelines of the Council of Bars and Law Societies of Europe, 'firms need to monitor the correspondence and communications of their fee-earners and other staff to ensure that their professional standards are maintained. If advice is given by staff by e-mail, firms will need to be able to check the accuracy of the advice' (CCBE, Council of Bars and Law Societies of Europe 2005, p. 10). To have an idea of the implications of this new market segment, the worldwide information management for content compliance market is forecast to pass the $20 billion mark in 2009, and grow at a 22% compound annual growth rate through 2005–2009 (IDC 2005) (Figure 16.6).

16.5 Future Trends: Towards the Legal Semantic Web?

Semantic Web technologies are currently applicable in a variety of law and e-justice domains. Drafting, sentencing, arguing, reasoning, multi-agent systems, and XML legislation are the current fields that AI and law researchers are developing.[6] But there

---

[5] In addition to these pieces of legislation, and following a series of corporate scandals in 2002 (namely, the Enron case), the US Congress adopted the Sarbanes–Oxley Act, which, among many other things, requires companies listed on US stock exchanges to comply with governance rules regarding confidential information. This requirement applies to European companies whose shares are traded in US stock exchanges, and also applies to European subsidiaries of US companies listed on US stock exchanges.
[6] See an updated relation of the ongoing EU projects and the state of the art in the field in Casanovas et al. (2007) and in Casanovas and Noriega (2007).

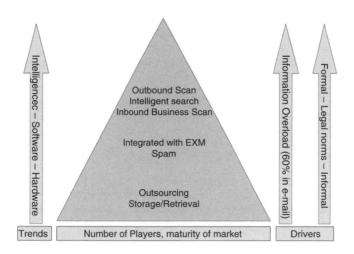

Figure 16.6   The EMM market

are other areas of e-justice that constitute a privileged domain for Semantic Web applications. Among these are electronic data interchange (e-filing, creation and retention and classification of electronic data to provide better, domain specific search engine capabilities), cataloguing and information retrieval at a particular database, web site or archive, knowledge sharing, and intelligent frequently asked questions systems.Multimedia and AI content technologies in law seem to be a promising field, not fully explored yet.[7] The same holds for audiovisual data integration, whereby data in various locations and various formats can be integrated in one single application. This is the case of e-Sentencias project in Spain.[8]

However, those are research trends, whose results are being absorbed slowly by the legal market. Accordingly, firms and companies are slowly adding semantics to other kinds of solutions for concrete problems, e.g. search and browsing, and meta-search engines (Sancho-Ferrer *et al.* 2007).

References

American Bar Association (ABA), 2007. Legal Technology Survey Report Results. Available at: http://www.abanet.org/tech/ltrc/survstat.html [accessed 10 September 2007].
Baldwin, T., 2007. *Is KM dead?* [Online]. Available at: http://kmpipeline.blogspot.com/2007/08/is-km-dead.html [accessed 10 September 2007].
Blanco, V., Martínez, M.M., 2007. Spain ongoing legislative XML projects. In C. Biagioli, E. Francesconi, G. Sartor, eds. *Proceedings of the V Legislative XML Workshop*. Florence: European Press Academic Publishing: 23–38.

---

[7] E.g. see the EU project Cognitive Level Annotation using Latent Statistical Structure, http://class.inrialpes.fr/.
[8] E-Sentencias (E-Sentencias. *Plataforma hardware-software de aceleración del proceso de generación y gestión de conocimiento e imágenes para la justicia*) is a project funded by the Spanish Ministry of Industry, Tourism, and Commerce (FIT-350101-2006-26).

Casanovas, P., 2007. *Use case: Helping new judges answer complex legal questions*, W3C, Semantic Web Use Cases and Case Studies [Online]. Available at: http://www.w3.org/2001/sw/sweo/public/UseCases/Judges/ [accessed 10 September 2007].

Casanovas, P., Noriega, P., 2007. Introduction: legal knowledge, electronic agents, and the semantic web. In P. Casanovas, P. Noriega, D. Bourcier, F. Galindo, eds. *Trends in Legal Knowledge, the Semantic Web, and the Regulation of Social Agents Systems*. Florence: European Press Academic Publishing: 15–41.

Casanovas, P., Sartor G., Rubino R., Casellas, N. (eds.), 2007. Computable models of the law. Language, Dialogues, Games and Ontologies. Lecture Notes in Computer Sciences 4884, Berlin: Springer. (forthcoming)

CCBE, Council of Bars and Law Societies of Europe, 2005. Electronic Communication and the Internet: Guidance for Lawyers [Online]. Available at: http://www.ccbe.org [accessed 10 September 2007].

Conrad, J.G., Schilder, F., 2007. Opinion mining in legal blogs. Eleventh International Conference on Artificial Intelligence and Law, June 4–8, 2007, Stanford, California, pp. 231–236.

Consejo General de la Abogacía Española, 2006. Annual Report [Online]. Available at: http://www.cgae.es [accessed 10 September 2007].

Dineen, S., 2006. *Law in Business IT: Think before You Send* [Online]. Available at: http://www.legalit.net/PrintItem.asp?id=30595 [accessed 10 September 2007].

Gelman, J., 2004. Legal Publishing and Database Protection. [Online]. Available at: http://www.law.duke.edu/cspd/papers/legal.doc [accessed 10 September 2007].

IDC, 2005. Worldwide Outbound Content Compliance 2005–2009 Forecast and Analysis: Content Security Turns Inside Out [Online]. Available at: http://www.idc.com/getdoc.jsp?containerId=prUS20017105 [accessed 10 September 2007].

Ikens, L., 2007. The 2006 ABA Tech Report: trends in courtroom technology: has the picture changed?, LJN's *Legal Tech Newsletter*, January 2007. Available at: http://www.ljnonline.com/alm?lt [accessed 10 September 2007].

ILTA, 2007a. Technology Purchasing Survey [Online]. Available at: http://www.iltanet.org/pdf/2007PurchasingSurvey.pdf [accessed 10 September 2007].

ILTA, 2007b. Telecommunications Survey [Online]. Available at: http://www.iltanet.org/communications/pub_detail.aspx?nvID=000000011205&h4ID=000000872605 [accessed 14 October 2007].

Katsch, E., Wing, L., 2006. Ten Years of Online Dispute Resolution (ODR): Looking at the Past and Constructing the Future. *University of Toledo Law Review*, Vol. 38, pp. 19–45.

Kersey, M. 2006. E-Mail Management: Turning a Compulsory Task into a Competitive Advantage [Online]. Available at: http://www.line56.com/articles/default.asp?ArticleID=7826 [accessed 10 September 2007].

Ministry of Industry-Red.es, 2004. La microempresa española en la sociedad de la información [Online]. Available at: http://observatorio.red.es/estudios/comercio/index.html [Accessed 10 September 2007].

Murphy, B., 2007. Roundtable discussion: E-discovery [Online]. Available at: http://www.kmworld.com/Articles/ReadArticle.aspx?ArticleID=37331 [accessed 10 September 2007].

Newman, P.M., 2006. Make the most of courtroom technology. *The Legal Intelligencer*, February 21, 2006. Available at: http://www.law.com/jsp/law/sfb/lawArticleSFB.jsp?id=1140170710550 [accessed 10 September 2007].

OASIS LegalXML, 2007. [Online] Available at: http://www.legalxml.org [accessed 14 October 2007].

Palmer, S., 2006. If you can't beat 'em, train 'em: how lawyers conduct legal research, ABA Legal Technology Resource Center, Available at: http://www.abanet.org/tech/ltrc/publications/lia_training.html [accessed 10 September 2007].

Poblet, M., Vallbé, J.J., Casellas, N., Casanovas, P., 2007. Judges as IT users: the Iuriservice example. In A. Cerrillo, P. Fabra, eds. *E-justice: Using Information Communication in the Court System*. IGI Books, Hershey, PA (in press).

Sancho-Ferrer A., Mateo-Rivero J.M., Mesas-Garcia, A. 2007. Improvements of Recall and Precision in Wolters Kluwer-Spain Legal Search Engine. In P. Casanovas *et al.*, eds. *Computable Models of the Law*. LNAI 4884, Berlin: Springer Verlag. (forthcoming)

Smallwood, R., 2006. *Email archiving and management: from niche to core component: Part 1, KMWorld Magazine* [Online]. Available at: http:www.kmworld.com/Articles/PrintArticle.aspx?ArticleID=15409 [Accessed 10 September 2007].

Smallwood, R. 2007. E-Mail management comes of age, *KMWorld Magazine* [Online]. Available at: http://www.kmworld.com/Articles/PrintArticle.aspx?ArticleID=35774 [Accessed 10 September 2007].

Svengalis, K.F., 2007. Legal information: Globalization, conglomerates and competition monopoly or free market. Available at: www.rilawpress.com/AALL2007.ppt [accessed 14 October 2007].

Swartz, N., 2007. The use and misuse of information new rules for e-discovery. In ILTA, ed. *Records Management: Beyond the File Room*, pp. 24–26. Available at http://www.itanet.org/files/tbl_56Publications/PDF33/142/Records%Management.pdf [Accessed 14 October 2007].

Violino, B., 2007. Digital dialogue: firms connect with online collaboration and wireless tools [Online]. Available at: http://lawfirminc.law.com/display.php/file%3D/texts/0907/amlawtech [accessed 14 October 2007].

Part Four
Final Words

17

Over the Horizon

Ian Pearson

17.1 Social Change

Most of the big social changes that we expect in the next 20 years are already well known – the pension problem resulting from an ageing population, the cultural conflicts and welfare problems caused by increasing immigration, the need for more housing as we see more and more single person households, crippling political correctness enforced by a surveillance-enthusiastic state, coping with an increasingly obese and unhealthy population. These issues are already being thoroughly debated. Adding to this list, we will soon be discussing the inevitable inter-generational conflicts that will arise as older people make larger demands on a system being funded by a decreasing number of workers and how to stop remigration being a resource drain as marketable young people move to younger countries with lower taxes, and immigrant populations move back to their homelands for the same reasons. In other words, the debate will move on to how we prevent the UK and other countries, e.g. in Western Europe, from becoming retirement zones.

But not all social changes are problematic. We are getting far more tools to enable social entrepreneurs to improve the fabric of our society, to link people together, to address loneliness, and to improve social inclusivity. These solutions are just as important to discuss as the problems we will face; otherwise, we can get too bleak a picture of the future. There will always be new problems, but as technology improves, so old problems also get solved. On balance, there is no reason to expect that the future will be worse than today, and in many ways, it will be better.

17.2 Personal Needs

Maslow documented human needs very well in his well-known hierarchy of needs. As a foundation for a happy life, we need to ensure our basic physical needs for survival,

ICT Futures: Delivering Pervasive, Real-time and Secure Services
Edited by Paul Warren, John Davies and David Brown
Chapter 17 © 2008 Ian Pearson

such as food, clothing, and shelter, to defend against predators and ensure our physical security. Once we have done those things, we want to belong to a society with other people, to strive to achieve some sort of status in that society, and to self-actualise. Most people in the UK can now satisfy the foundation layers of this hierarchy relatively easily, and the expanding markets are at the top of this pyramid now. The upper layer markets absorb an increasing share of our income each year, and we should expect that to continue. We will spend more of our funds on status, socialising and trying to achieve more with our lives, to do more interesting things. Technology increases the number of options for us at these upper layers. It becomes easier to network and to find new friends using web based tools, and also easier to find out what is available to us as ways to spend our time.

The trouble with this is that with so many things available to us, it becomes harder to make decisions. As a result, we feel that we could be missing out on something even more interesting. The grass is always greener . . . So instead of increased wealth and choice of lifestyle making us happier, it can actually make us feel more stressed and confused. Not knowing how to best spend the limited time we have undermines our basic feeling of security and one of the foundations of happiness. At the same time, the diversity of choice of ways to spend our time helps to fragment society. People make different choices, so they have less in common with their neighbours. Socialising with people further away, and forming tribal allegiances with geographically distributed people undermines the strength of our local geographic community.

However, community networks are springing up everywhere, and this offsets some of the damage. It is easier to find people of like mind, but also easier to do so in the nearby area, now that most people are on the net. Some innovative social groups such as Freecycle (http://www.freecycle.org/) help to forge links between local people by latching on to their common desire to reduce waste, while simultaneously helping less well-off people. People get something they need for free and the donors get a nice warm feeling that they are 'good people'. The fabric of society is a little stronger as a result. We will see much more of this kind of social innovation as people get more used to using the web.

The future will see many of these positive and negative forces acting on society, via the individual decisions we all make, through a myriad different motivations.

17.3 Lifestyle Change

There are lots of lifestyle choices available to people, but most people want to do pretty much the same things, since their aspiration are generally motivated by the same social pressures. Wanting to belong and to conform mean that we can have a wide variety of cultures around the world, but much less variety locally. When social trends happen, they tend to happen to large numbers rather than to just a few. It is interesting to observe that the latent need for a trend can build up enormously before it actually starts to happen. The recent explosion in the numbers of people using the web is a good example. With just a few people using the web, there was little incentive for other people to join in, but once most of their friends were also using it, there was huge pressure to do so, and the web suddenly took off. The dotcom crash was caused at least in part by the lack of critical mass. Not enough people were on the net. Now, almost everyone uses it regularly, so critical mass is there. Many businesses that collapsed in 2000 will now succeed well.

In the 1960s, the hippy culture was very significant. Many of the same social foundations and pressure exist again now. There is once again a strong emphasis on emotion: love for one's fellow man, new age thinking, a disrespect for authority, anti-war, and so on. There are a few people around who are already behaving like hippies again, but not many. But the pressure is building and it is highly likely that we will see a strong resurgence of nouveaux hippies, once the right celebrities give people the sign that it is now fashionable again and OK to join in.

17.4 Messaging

Messaging is simply asynchronous communication. Instead of interrupting each other to chat in real time, messaging allows the other person to deal with the communication in their own time. It is extremely useful in today's time-scarce lifestyles. That sometimes the messaging becomes essentially a conversation does not change its usefulness. But there are some future developments that will make messaging even more useful than today.

Today's messaging works well between people who know each other already. It is easy to message friends and colleagues. What is missing is the ability to send messages to interesting strangers. Being able to send a message to someone you fancy just by pointing your mobile phone at them would be a big breakthrough, as would inter-car signalling to talk to the person next to you at a traffic light, or the guy hogging the lane in front of you. Also, people will want the ability to leave messages in a particular location for other people who will pass through that same space, comments about restaurants, for example. This requires some sort of social structure before it can work well. There needs to be a filtering process, allowing people to leave genuinely useful comments, while preventing junk marketing, or it won't work.

Digital bubbles can achieve exactly that. Electronic jewellery will radiate public information about the wearer into the nearby space. People passing them will have electronic access to that information, and if it is interesting to them, they would see the relevant bits in their head-up display or hear it in their earpieces. Physical objects, shops, and vehicles can use similar technology. We will therefore be bathed continuously in an intense information field. Our digital bubbles will act as a semi-permeable digital force field. Our computers will know us very well, know what we like, where we are going, with whom, why, and so on. They will use this context information to filter out the trickle of information that we would want and give it to us, ignoring the rest of the available information.

This kind of passive messaging between digital bubbles is likely to affect many of our interactions. Networking would be greatly enhanced if our computers have already worked out whether someone is likely to be of interest to us before we even get to them. We would meet both useful business contacts and new friends, with far less of the time wastage involved in the preamble of conventional social interaction.

17.5 Anti-tech Backlash

Of course, all of this Increased use of technology will leave a bigger digital trail, allowing far more electronic surveillance. Privacy reduction is one of the costs of improved connectivity. The system needs to know who is where and what they are doing in order to make these kinds of useful links, but that means we have to sacrifice some anonymity. When the system is run by a trusted organisation, then there is little problem. If it is run or monitored by government, the police, or other parties with another agenda, such as insurance companies, then we are likely to question the privacy trade-off much more. In fact, there is already a great deal of media discussion of the loss of privacy in today's society, with warnings even from the Information Commissioner that we may be sliding towards a surveillance society.

In parallel with this increasing surveillance, people are increasingly aware of globalisation and its consequences for jobs and their personal wealth. They understand that much of this change is enabled by improved communications and other information technology. Over the next decade or two, we will see huge increases in the effective intelligence level of our computers, bringing them to the point where a great many of today's information economy jobs can be automated. For example, the Semantic Web, described in Chapter 5, will allow ready automation of many administrative, data searching, processing, and even knowledge creation jobs. Many people will have to re-train, some several times in a lifetime, as technology evolves. The far future will provide jobs that are mainly to do with human skills rather than with intellectual or manual skills, and many people will not feel comfortable in that world.

With parallel threats to livelihood and personal freedom, both associated with new technology, there is likely to be some kind of backlash against technology. This will happen in an era where people will be learning to use the net to mobilise political activities. It is likely that an anti-technology/anti-globalisation movement will be one of the web's first major political experiences.

17.6 Political Trends

Political activities are already making some use of the web. The Burma demonstrations were accompanied by web activity, too, as have been some anti-globalisation demos in London, and various electoral campaigns. The web is now a permanent part of the political platform. As the web becomes more ingrained in every activity in normal life, so it will become a more significant platform for political battles. The web has some major advantages over conventional platforms such as the soap box in the park or the TV or radio, or newspaper front page. It is far more instant than any of these, and has the enormous advantage of not being tied to any particular geography or time zone. While some politics obviously involve local issues and so will remain locally debated, some ideological issues are much less geographically linked, such as feminism, poverty, environmental activity, etc. These are much better suited to web activity. Being able to send messages simultaneously to large groups anywhere in the world is a major advantage. The same networks can then be used to link people's personal computers together to orchestrate follow-up activities such as electronic attacks. It is also much easier to police membership of organisations on the web than physically. People can be evicted from web groups easily.

The world is becoming much more globally interconnected. Activities in one country that affect another are being addressed more now, compared with the 20th century where politics largely stopped at a country's borders. So we are now seeing campaigns by African countries seeking damages from rich countries because of the environmental impact of our previous activity. I am pleased to report that we predicted that in the early 1990s! The world is now a very different place. In parallel with this, there will also be more campaigns in developed countries to help those in less developed countries.

Such trends are likely to increase over the coming years. The better we are connected with people in other countries, so the more we feel they are part of our community and respond to their needs better. Thanks to telecoms, the world seems a smaller place. Improving the global sense of human community is a wonderful by-product of providing platforms for global business.

17.7 Needs of Groups

Telecoms used to be a two-domain market – business and personal. It is already more complex. As a third domain, it is now obvious that the network and IT infrastructure in itself also produces lots of data that need to be networked, just for the ongoing working of that infrastructure, and to fulfil its primary purposes such as collecting and distributing data about the environment and itself. Furthermore, social groups need communications services. They always did, but they were not available across networks, so they were implemented by primitive paper-based systems like newsletters. Most group activity relied on face-to-face communication. Now, as people are spending far more time online and have mobile devices in their pockets all day, they use these to keep in touch with activities in the various groups they subscribe to. Teenagers keep in touch constantly with their friends, and clubs and societies have their echoes in chat rooms and networking sites that also allow messages to get to their members wherever they are.

This marketplace is beginning its next phase of increasing complexity with the fifth domain, the virtual world. Actually, it permeates all the others, allowing convergence of each physical world market with its online part. So now we have avatars in virtual worlds making voice over IP calls to each other. Of course, the participants are usually people, but they need not be, and in the future, we will see more chatbots and artificial intelligence (AI) entities taking part in many social and business activities, generating traffic just as humans do. The same applies to virtual infrastructure. A business is just as interested in the comings and goings to a virtual shop as to a real one.

Groups need more than just one-to-one communication. Often, people want to communicate with several people at once, and this subset of their overall contact lists might vary from communication to communication, so we need easy ways of talking to subgroups. With people carrying different equipments, we need more services to translate between media, such as voice to text or text to voice translation. Groups of friends would welcome new tools such as being able to see where their friends are just by glancing at their mobile phone display. This enables them to make more use of occasions where they chance to be nearby one of their friends with some free time. Today, many such social opportunities are simply lost because we simply don't bother to track every one of our friends all of the time.

Of course, some groups exist to support the specific interests of that group. Political groups would expect to make good use of the real-time communication opportunities afforded them to help enhance their political power.

17.8 Care Economy

We are supposedly in the middle of the information economy. Knowledge and its use is the primary source of wealth generation. However, like the agricultural and industrial economies before, the information economy will not last forever. I have already noted how new tools, such as those of the Semantic Web, are helping to automate a great many administrative tasks. Not long after, AI will approach human capability levels across a wide range of activity, and the associated jobs will also succumb to automation. As the price of information work falls, so its contribution to the economy falls. It will still be an essential foundation, but we just won't have to pay much for it, so it will no longer be a key part of the economy. Instead, as we always do, we will place the main importance on what people do. That is what we always have to pay for. When physical work can be done better and cheaper by robots, and information work better and cheaper by AI, people will be forced to concentrate on those activities that can still be best served by people. Essentially, this means that interpersonal, emotional, human skills will dominate. This is called the care economy for lack of a better term, because caring skills represent the biggest part of it. So, while a hospital consultant could notionally be automated by an expert system in a sophisticated robot, a nurse will still be human because a nurse's caring skills depend essentially on being the same species as the patient to maximise empathy. A robot will not be as good, even if it has a cute android body and pleasant synthetic personality. The care economy is not entirely new. Twenty-five percent of today's population work in care economy jobs such as care, policing, teaching, and personal services. But these jobs will grow in number in the same way as the service economy grew in the second half of the 20th century.

17.9 Globalisation, Localisation

Globalisation will continue for at least another decade, but as the care economy starts to take hold, we will value face-to-face contact far more than the mechanistic side of transactional processing. It will be much less appropriate to outsource work to faraway countries. Globalisation of business practices, military, environmental and legal foundations will probably continue, but we probably will see a relocalization of work.

Similarly, while we see a globalisation of many aspects of politics, we will also see more focus on using community networks to organise local politics. As basic platforms homogenise around the world, dictating fairly standardised human rights, medical care, education, and so on, so national level governments will have far less to do. Although we should not expect a single world government, we will certainly see increased global regulation, with local decisions being taken very locally. So, for example, the location of a school might be decided by the local community, but the existence of the school is dictated globally. National government will gradually evaporate.

17.10 Virtual Business

The web encourages fragmentation and reassembly of businesses and organisations. Web-enabled organisations often have a different form from those that preceded them. A major trend now is for experts to ask themselves why they need to bother working for a company, when the web allows them to work freelance much more easily and still make a good living. In some sectors, this is causing a big problem for large companies, who cannot attract the staff they need in some roles because those staff can work for the highest bidder on a day to day basis. It is amusing that the same technology that has allowed blue-chip boards to downsize and give higher shareholder returns via automation has also allowed automation of the company itself. Staff are asking why they need the company! The original purpose of the company was to provide an organisational and resource structure that allowed large market niches to be addressed, and this can now often be done quite easily by groups of freelancers. Web applications that can link freelancers together into virtual businesses allow many market sectors to be addressed better by virtual business than conventional business. Without the overheads of expensive layers of redundant management, virtual business can compete easily and often win. As web tools get better and as staff learn how to use them, and gain the confidence to leave the security of the corporation, many such corporations will lose their market edge and die.

17.11 IT Renaissance

We are currently in something of a dark age as far as IT practices are concerned. In the early 1990s, IT was the exclusive domain of IT people. We saw the creation of the web, email, messaging, web sites, mobile phones, PCs, and so on. Accountants gradually learned the usefulness of these tools and started to automate finance within companies, from online expense claims to e-commerce. This worked fine too, mostly. Finally, administrators discovered IT, and we dived into a world where every manager could suddenly send missives and directives to everyone in the company to indulge every idea that crossed their mind. Since it was fast and easy, and there was always some headline saving or improvement, no one ever seemed to count the company-wide costs of doing so. The result today is that almost every large company has a huge number of ill-informed practices that almost all fall into the categories of micromanagement and centralisation. It is easy to implement a practice that might cost 100 times more than it supposedly saves, once the end-user costs of lost productivity are properly taken into account. This happens because most companies have no way of measuring the minute by minute output of their employees, so these costs are simply ignored. After a number of such practices are implemented, a severe drop in the productivity of the company might be experienced, without anyone knowing or even noticing, since it happens so gradually.

However, although company managers might be fooled occasionally, markets are never fooled for long. New companies entering the market that don't use such poor practice will win market share. Gradually, bad practices will be filtered out of the market, either by simple extinction of the unfit, or by badly run companies realising their errors and entering the age of IT enlightenment. This new age, the IT renaissance, where IT gets the 30% efficiency gains it always promised, but without throwing those gains away on some other bad practices, will finally see the proper rewards of IT being realised. IT will be mature, and managers will no longer misuse it to the same degree. IT management will have grown up.

17.12 Socially Integrated Companies

In the care economy, most businesses will have to act on a much more local basis to survive. Human contact will dominate the quality of relationships with customers, and those companies best suited for survival will be those that integrate best into the local community. Emphasis on quality of life is already increasing rapidly, as people realise everywhere that money is only one factor among many that dictate quality of life. So, we will see ongoing blurring of the boundaries between home and work life that is already happening today, and with diffusion of the borders of companies as they move increasingly into the virtual, so the boundaries between the community and the organisational community will blur too.

17.13 Urban Development

Urban development today is seeing the deployment of ubiquitous wireless local area networks, followed soon by a fully fledged ambient intelligence environment, with sensory networks, storage, and communications seamlessly interwoven into the city infrastructure. In parallel, personal interfaces to IT will soon include head-up displays, overlaying computer generated data and images into our field of view. This will enable a convergence of the real and virtual worlds, as web functionality starts to overlay the city streets via these head-up displays. When we live in a dual world, with physically real objects mixed with virtual ones, we can start to redesign the city environment for both the virtual and real worlds, making full use of positioning and recognition technology to align virtual imagery with physical objects, people, and buildings. Although there can only be one physical appearance, an object can have any number of virtual appearances, which would depend on the person looking at it and the context of the situation. Shops can be designed with a multitude of virtual layers and appearances, the duality producing a much bigger virtual marketplace than would physically fit into the same area. Using duality to the full will make the urban environment much more enticing and potentially much richer culturally too.

17.14 Simplicity

Although people want the rich functionality afforded by modern technology, most people find it hard to cope with the enormous complexity of the innumerable interfaces around us. Most things are far too complex, such that when a company develops something that is attractive and yet easy to use, we are surprised. In the future, we should expect much more emphasis on trying to achieve simplicity.

One area that will be addressed is the computer. Today, most of us use PCs. They have extremely complex operating systems to cope with a wide variety of tasks, all of which are scheduled to run on a single processor, or at best two or four. With the same chip fabrication technology as we already use, it would be possible to make computers that run with thousands of processors, each of which is far smaller, cheaper, and simpler, and to throw away most of the operating system. This would give many advantages to the computer designer, increasing reliability and speed, reducing cost, and allowing much simpler resource management. In doing so, and moving back once again towards encapsulating software on memory chips rather than DVDs, it would also be possible to eradicate most of the routes used by hackers to cause security problems for us. With less need to run continuous virus protection and firewall software, the operating system could be made simpler still. At each of these advances, and many other potential modifications, computers would become simpler, cheaper, more robust, less error prone, and more secure. It is hard to see any disadvantages of doing so. We might consider that we have been pursuing the wrong path to serving our computer needs for the last 25 years as this new route to computer design is explored. But it is not too late to change path now, and we should expect that we will.

17.15 Storage Technology

Storage technology is one of the great success stories from Moore's Law. Memory sticks have come in the last few years from being a novelty to being able to hold all of the files that we need to do our jobs. For example, I can fit all of my personal output since 1985 on my memory stick. This creates new options for the ways in which we work. It means we are no longer stuck to a PC on our desk, but are now free to work from anywhere there is a computer.

Memory sticks also provide the promise of a new distribution route for information and content. We are a short hop from making our memory sticks wireless, and when that happens, it will not be long before they can talk to each other. Just by walking past someone in the street or rail station, public area files will jump between the sticks, according to our personal preferences. Instead of having to log on to a web site to get a podcast, we may get it automatically from some stranger we pass on the way to work.

17.16 Machine Consciousness

The technologies of nanotechnology, biotechnology, and information technology are already converging. Gene chips are already able to detect individual molecules, for example. Computers already use nanotechnology components, and we see the use of

IT in medical implants routinely. Another wave of convergence will be where all of this meets our nervous systems, which are essentially just organic IT. Once we can link routinely with nerve endings, we can begin to record and replay sensations, making multisensory communication environments, adding sensations to computer games and online shopping and the like. In a few decades, we will be able to augment human memory, sensory, and thinking processes, at first for alleviation of senility, and later for purely cosmetic or enhancement reasons. It is a short step philosophically from there to making a full direct brain link to back us up in case of accident, and achieve electronic immortality. Realistically, this is likely around 2050 for a few, and 2070 for anyone. Already, neuroscientists are progressing with reverse engineering the human brain, learning how it works and encapsulating that knowledge into computer designs. The hippocampus can now be replaced by a synthetic one, at least in rats, and actually improved memory storage compared to nature. The coming years will bring an accelerating return as better sensors and computers and interfaces lead to an acceleration of neuroscience and biotech, with the positive feedback being completed by the consequent insights being used to accelerate computer design. Some scientists believe we will routinely have electronic implants connected to our nervous systems by 2030, and the natural progression of such a trend leads to the situation by 2050 for many people where almost all of their mental capability resides in the machine rather than their biological components. Thereafter, biological death will no longer mean the end of our minds. Mentally at least, we will have the potential for immortality.

However, this human progress will not happen independently of progress elsewhere in the machine world. Today, even the best machines are well below human capability in terms of thinking, and no machine yet is conscious in the human sense of the word, however well they might play chess or imitate chat. This will change. With the assistance of nanotechnology and improving scanning techniques, neuroscientists are beginning to reverse engineer the human brain. With each insight into how the brain works, the capability to produce a conscious machine comes closer. A conscious machine does not have to work exactly the same way as our brain, of course. There must be many ways of achieving consciousness, and we only need a few starting ideas to give to a computer before it can use evolutionary techniques to experiment and accelerate development towards that goal. It is probable that we will not understand how our machines work in the future, any more than we really understand how our own brains work today, but there will probably come a point some time between 2015 and 2020 (where we will see the first machines that demonstrate self-awareness and consciousness in just the same sense that we do – several scientists around the world are already considering approaches that are likely to achieve this, and are just waiting for the platform technology to develop according to Moore's Law. They will almost certainly not be like us, and they will think very differently, but their alien consciousness will be just as real as ours.

However, we can produce electronic devices that send signals many times faster than signals travel in our brains, and switches that operate far faster than our nerve cells. It is likely that these first conscious machines will use electronic and optical technologies that are fundamentally far faster than our organic ones. With so many spare zeros to play with, even if their algorithms are crude, there is a high chance that they could be superior in overall intelligence terms. After a few years of development, itself done mostly by smart machines, we should expect machines that far exceed human intelligence.

So as we develop links between our nervous systems and the machine world, we will be connecting to a world with much greater capability than nature gave us. With the potential for superior senses, superior intelligence, and memory, and the obvious derivative of telepathic linking via the networks to other similarly equipped people and machines, electronically enhanced humans will be fundamentally different from natural ones. This development is likely by default over the next few decades, but its implications are profound for humanity. It may be gradual, but it will not be taken without much vigorous debate, and possibly conflict. The big question is whether a smarter than man machine be called a productivity aid, or Sir.

18

Conclusions

Paul Warren, John Davies and David Brown

18.1 Fundamental Trends

Our book has taken a broad sweep across the whole range of information and communication technologies. We have looked at the technologies themselves, at how they are influencing society, and in turn how societal demands are influencing the evolution of these technologies.

How can we summarise all these changes? What are the major trends which characterise the direction of our industry?

We suggest that there are five fundamental such trends. Three are technological: the move towards delivering applications as services, the deployment of semantic technology, and the ubiquity and pervasiveness of our technology. One is organisational: the increasing emphasis in business on collaboration. The final one is social: the net as a place for social interaction. Sections 18.2–18.6 discuss all these in more detail. They all have implications for security, and for this reason, security is discussed in Section 18.7. Finally, in Section 18.8, we draw our book to a close with some last words on what we hope we have achieved.

18.2 Delivering Services

Throughout our industry, there is an emphasis on 'service'. What does this mean? The term service economy has been used for many years to denote economic activity focussed on intangible goods rather than on manufactured goods. Here it means something more specific. When we use terms like software-as-a-service (SaaS), service oriented architecture (SOA) and service oriented infrastructure (SOI), we are firstly saying something about *ownership*. We are saying that increasingly, the beneficiary of a piece of technology no longer needs to own it.

ICT Futures: Delivering Pervasive, Real-time and Secure Services
Edited by Paul Warren, John Davies and David Brown
© 2008 John Wiley & Sons, Ltd

This is of particular advantage to the individual and the smaller organisation who may not wish, or may not be able to afford, to purchase software licences. So, Google offers Web-based office applications to the individual, whilst salesforce.com offers enterprise applications to organisations. Even large organisations benefit. The corporate data centre can be dimensioned for normal usage, and distributed computing technology used to make resources available for peaks of demand from elsewhere, or to give access to specialist resources very occasionally required.

This change in charging model frequently damages the supplier's margins, but competition gives him no choice. In any case, the supplier also benefits as new markets become available; where customers could not afford to purchase products, they could purchase services.

SaaS and SOI are not just about cost of purchase. Maintenance costs are reduced because maintenance is done centrally. Neither bug fixes nor software enhancements need to be distributed. There is a real saving here; for even when done over the net, there are still organisational costs to maintenance and enhancement.

Even within an organisation, the application of SOA is about a shift of ownership. The application owner no longer needs to own all the components which enable the application. Some are owned centrally or by other departments, and are called when needed at run-time.

Service delivery is also about *reuse*. Reuse has been central to the development of computing. The replacement of machine code programming with high level languages was a triumph for reuse. The programmer who writes '$x : = x + y$' is reusing a machine code routine to achieve this addition. Subroutines and subroutine libraries took this one stage further, but the reuse is still at compile time. With SOA and Web services, the reuse is at run-time, and we are potentially talking about reuse across the global net. The promise for human productivity is enormous.

The service philosophy is not just about sharing hardware and software. Data, information and even knowledge can be shared. Data banks, e.g. in the biological sciences, can be accessed remotely when needed; information can be accessed on-demand, e.g. as more and more journals and even books are available. Knowledge is also available, as experts can be located and consulted over the Web.

18.3 Semantic Technologies

Implementing the service philosophy across a global Web requires that the resources and services on the Web be semantically described. At the global scale, we need to be able to find the services we want out of hundreds of thousands or more available, and we need to compose them automatically. This can only be done when they are described formally such that a machine can make inferences ('reason') about them. This is true whether the services provide access to ICT infrastructure, to software or to information.

Apart from its role in enabling our service oriented vision, the use of semantic technology is essential to enable the highest level of interoperability within and between companies. The need to create a unified and consistent view of corporate data is crucial in an era when mergers and acquisitions require the rapid coalescing of organisations, and when partnerships form rapidly, to create supply chains, to undertake innovation and generally to achieve business value.

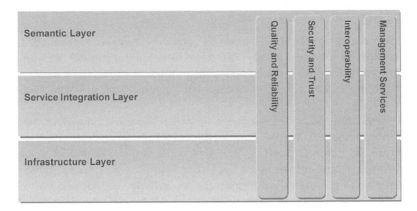

Figure 18.1   The Networked European Software and Services Initiative vision for services and software

In addition, semantic technology is the key to unlocking the great mass of organi-sational information which resides in uncodified form, in reports, emails, on intranet pages and scattered everywhere across an organisation. Apart from the inherent busi-ness value of this information, knowing what you know is essential to organisations facing legal disclosure challenges, e.g. the financial disclosure required by the US Sar-banes–Oxley Act.

The Networked European Software and Services Initiative[1], which exists to provide a unified view for research in services architecture and software infrastructures, has illustrated the central role of semantic technology in the diagram reproduced here as Figure 18.1 and in the following paragraph, both taken from its vision document (NESSI 2005):

> Semantics will be a key element for the transformation of information to knowledge. One way to build knowledge will be through advanced search engines that allow fast search in large unstructured data. Semantic Web technology based on ontologies will enable far more effective machine to machine communication about the nature and the manipulation of data they hold and actions based upon that data. On the business process level, business modelling provides the semantics that is required for business process management, process transformation and inter-company cooperation.

Semantic technologies require metadata; that metadata will come from various sources. Chapter 5 discusses the use of automatic or semi-automatic techniques for creating metadata. With the advent of folksonomies, informal tagging has created a rich new source of metadata. A current research challenge is how to get the best out of a combination of the formal and informal approaches.

Chapter 12 made clear the real practical progress which the discipline of artificial intelligence has made. The combination of semantic technologies, the increasing avail-ability of metadata and techniques from artificial intelligence will enable powerful applications both in the enterprise and on the global Web.

---

[1] http://www.nessi-europe.com/Nessi/.

18.4 Ubiquity and Pervasiveness

The pervasiveness of our technology can be interpreted on several levels. Chapters 3 and 11 talked on the physical level, describing how our technology can be physically omnipresent. Chapters 13–16 described a different sort of pervasiveness; illustrating how our technology pervades so many industry sectors and so many aspects of our lives.

Physical pervasiveness itself has a range of forms. We are used already to being able to do computing wherever we go. Usually, this is by taking our computing devices with us, but Chapter 17 has reminded us that this is not really necessary; we could carry our computing environments on a memory stick and plug into an available computer. We are used also to the idea that computers be embedded in all sorts of other appliances, from washing machines to weapons of war. Hitherto, this has been at the level of a few processors per device. We are now moving to a world where thousands of sensors can easily be deployed in a given environment, and where computational intelligence can be deposited onto physical components in a manner akin to printing. Chapter 8 explained how high bandwidths can be made available for these devices. Often, though, applications will not require particularly high bandwidths. Peer-to-peer radio networks, offering greater flexibility, will frequently suffice. This creates from the physical world a parallel world on the net. The interaction of these worlds will enable rich applications over the coming years. For a discussion of this, complementary to that given in Chapter 11, see Warren (2004).

ICT's pervasion of our working and personal lives will change how we view these technologies. The doctor, for example, will no longer see ICT as something different from medicine, introduced to solve certain specific problems, but as providing a set of tools as fundamental to his profession as the stethoscope. As the creation and management of ICT infrastructure becomes more and more a commodity skill, the real professional advantage will come to those who are thoroughly skilled in their own worlds and who also really understand how to use ICT to create value in those worlds. These are the people who will innovate to create future leading edge applications.

The trends to service delivery, deployment of semantic techniques and pervasive ICT do not exist in isolation. Chapter 5 described how semantic technology could be used to characterise services. This can be extended so that semantic techniques can be used to characterise any ICT resource, and indeed, any resource can be made available as a service. This is true of traditional ICT resources; it is also true of sensors and actuators through which the world of information interacts with the physical world. For a vision of how semantic technology can be applied to both conventional computing devices and also pervasive technologies, and the kinds of applications which will ensue, see de Roure *et al.* (2005).

18.5 Collaborating in an Open World

The success of the Open Source movement has alerted mainstream opinion to the existence of a phenomenon which is older; that innovation does not just come out of research laboratories surrounded by security fences, but frequently arises when people work together across organisational boundaries. Innovators do not, by any means, all work for large organisations, and many of them do not work for organisations at all. Some of them do not even work for profit, in the normal commercial sense of the word. In any case, the world many innovators inhabit has, by necessity, to be much more open than the conventional world of the company research laboratory.

von Hippel (2005) talks of *lead users* who work with suppliers to create innovation. He makes us aware that this phenomenon extends well beyond ICT, and in referencing Allen's (1983) work on the 19th century English iron industry, he reminds us that it is by no means a new phenomenon. Of course, ICT acts as an enabler for the sharing of knowledge which this way of working requires, whilst the purely intellectual nature of, for example, software development makes it a natural target for collaborative working.

Tapscott and Williams (2006), drawing inspiration particularly from the phenomenon of Wikipedia[2], see collaboration as a major force for future innovation. Collaboration can be between organisations and their customers, the 'lead users' of von Hippel; between organisations and their suppliers; and between the suppliers themselves. It can also include private individuals, outside of any organisation, as in the case of Wikipedia. It is ICT which creates the low barriers to entry to make this possible. Not that Tapscott and Williams see the company R & D laboratory as being completely redundant. In their view, the latter still has a significant role, as part of a richer innovation ecosystem.

Whether inside or outside an organisation, collaboration ultimately takes place between individuals, and Chapter 2 discussed how tools such as email and instant messaging contribute to personal productivity, and how they also have drawbacks. More recently, wikis and tools utilising the informal tagging of content have offered benefits to complement and supplement the more traditional approach to communication and information sharing. Tapscott and Williams (2006) draw attention to the take-up of wikis in companies like Dresdner Kleinwort and Xerox; the editors of this book can testify to their similar use in BT. Better tools will continue to emerge to help us walk the fine line between ignorance of what we need to know and being slowed down or overwhelmed by less relevant information.

18.6 The Net as a Meeting Place

The last section touched on the techniques of social computing, e.g. the use of informal tags to enable the sharing of information. These techniques have largely been adopted from a world of private individuals, using computers for pleasure rather than to achieve organisational goals. In this world, the Web is being used by private individuals not just to achieve practical ends, such as booking a flight or researching an illness; it is being used as a place for social interaction for its own sake. Using the Web is not just about interacting *with* the Web. It is also about interacting with one's fellow human beings *through* the Web. The Web has become a meeting place.

---

[2] http://www.wikipedia.org/.

This phenomenon includes online chat rooms, the sharing of photos through flickr,[3] the popularity of Second Life[4] and the rise of Facebook[5] as a way of communicating between friends. Of course, we shouldn't have been surprised. Chapter 1 reminded us that the telephone was not conceived originally for person-to-person communication, and that examples are not uncommon of technologies devised for one set of purposes, often utilitarian and often unrelated to social interaction, being adopted by people to enrich their social interaction.

What this does reflect is that when you give people tools, they will use them for whatever matters to them, and that often means communicating with other people.

18.7 Security

In organisations, although perhaps not in our private lives, security is not a primary requirement. Organisations do not exist to be secure; they exist to achieve organisational goals. Nor do they primarily want their ICT to provide security, except in special cases such as systems to monitor and control physical access. They want their ICT to provide functionality, and then they want that functionality to be secure. This is not to say that security should be an afterthought; it should be designed into a system from the start. It does explain, though, why security often is an afterthought.

Reflecting on the themes described in this chapter, we see that they all place new challenges on security. An SOI spanning different organisations requires the kind of security control discussed in Chapter 7. The reliance on a Semantic Web to provide information means that we must be confident of the provenance of that information. It is for this reason that digital signatures were included in the Semantic Web's original architecture; see Figure 5.4. The ubiquity of our technology, e.g. the large-scale deployment of RFID tags, creates threats in response to which we need sophisticated security technology and processes. Similarly, the pervasiveness of our technology through so many aspects of our lives, e.g. our health and finances, makes security an even greater imperative. Collaboration and sharing are only possible if we are confident that only what we want to share is being shared, and only with those with whom we wish to share.

Finally, the rise of the net as a meeting place for private individuals has, as a disadvantage, the possibility of identity theft; of abuse of the innocent or naïve, particularly the young; and of providing a rich source of data for anyone with evil intent against the individual. To combat all these, we need vigilance, but that vigilance needs to be augmented with technology.

18.8 ICT Trends – Prediction and Use

'Prediction is very difficult, especially about the future'.

This quote, commonly attributed to the physicist Niels Bohr, sums up the challenge of our book. Not everything we have predicted will prove correct, although we believe

---

[3] http://www.flickr.com/.
[4] http://secondlife.com/.
[5] http://www.facebook.com/.

that the major trends we have identified will hold true and be significant, at least for the next decade. More importantly, we hope that this book will have stimulated readers working in disciplines beyond ICT to think about how the evolution of ICT can be used to benefit their own fields of endeavour; and for those readers who are our colleagues within the domain of ICT, we hope that our book will stimulate them to think holistically about how their work can influence, and be influenced by, their co-workers. In achieving this, the book will have achieved its purpose.

References

Allen, R.C. 1983 Collective Invention. *Journal of Economic Behaviour and Organization* 4(1): 1–24.

de Roure, D., Jennings, N.R., Shadbolt, N.R. 2005 The Semantic Grid: Past, Present and Future. *Proceedings of the IEEE* 93(3): 669–681.

NESSI 2005. NESSI Vision Document, Network European Software & Services Initiative.

Tapscott, D., Williams, A.D. 2006 Wikinomics. Penguin Group (USA) and Atlantic Books (UK).

von Hippel, E. 2005 Democratizing Innovation. The MIT Press.

Warren, P.W. 2004 From Ubiquitous Computing to Ubiquitous Intelligence. *BT Technology Journal*, 22(2), 28–38.

Index

ICT Futures: Delivering Pervasive, Real-time and Secure Services
Edited by Paul Warren, John Davies and David Brown
© 2008 John Wiley & Sons, Ltd